3/13

A Palette of Particles

A Palette of Particles

Jeremy Bernstein

The Belknap Press of Harvard University Press

Cambridge, Massachusetts

London, England / 2013

Copyright © 2013 by
the President and Fellows of Harvard College
All rights reserved
Printed in the United States of America

Library of Congress Cataloging-in-Publication Data

Bernstein, Jeremy, 1929–
    A palette of particles / Jeremy Bernstein.
      pages   cm
    Includes index.
    ISBN 978-0-674-07251-0 (alk. paper)
  1. Particles (Nuclear physics)—Popular works.   I. Title.
QC793.26.B47   2013
539.7'2—dc23   2012032819

Book design by Dean Bornstein

Contents

A Palette of Particles

Introduction

I have been exposed to the physics of elementary particles for over a half century. These particles now appear to me as colors in a palette that can be used to compose the tableau of the universe. There are what I would call the primary colors: the particles needed to describe the aspects of the universe that are accessible without the use of tools such as very high-energy accelerators or cosmic ray detectors. In this list I include the electron, the quantum of light (known as the photon), the neutron, the proton, and, most exotically, the neutrino. I am also willing to include their antiparticles, although most of these are not accessible without advanced technology. This was the set that was in play until the 1930s. At this time a new set of particles appeared, either in the minds of theorists or unexpectedly in experiments. An example was what came to be called pi-mesons and later pions. They were a theoretical suggestion of a Japanese physicist named Hideki Yukawa, about whom we will hear later. Initially it was thought that they had been discovered in cosmic radiation—radiation from extraterrestrial sources—but it turned out that what had been discovered were not Yukawa's particles but mu-mesons (now called muons), which are heavy, unstable electrons. Before physicists could resolve the matter and resume the search for Yukawa's particle, however, World War II intervened.

Upon returning from the war, the physicists resumed their ordinary research, discovering a plethora of new, unexpectedly strange particles either in cosmic rays or in the collisions produced in the new, very high-energy accelerators. These included particles such as the K-meson and what might loosely be described as heavy neutrons and protons—particles such as the so-called lambda and sigma particles. They seemed to have no rhyme or reason until Murray Gell-Mann proposed an order and eventually the components that he called quarks. There are now dozens and dozens of these strange objects. Lastly, there is the Higgs boson, the existence of which appears to be confirmed in the experiments being done at CERN's Large Hadron Collider near Geneva, Switzerland. This particle is thought to be the source of the masses of at least some of the others.

This chronology explains the order of the book. I begin with chapters on the primary colors. Then I turn to the secondaries and finally to the exotics. I hope in each chapter to give some piquant bits of history. I was witness to the postwar developments, so I bring in an occasional personal observation. Part of the material in this book was previously published in *American Scientist,* and hence the level is attuned to a general reader with an interest in science.

Before I begin, let me make a general observation. None of the particles we shall discuss can be seen with a microscope, to say nothing of the naked eye. The quark, if present ideas are correct, can never be observed by any means as a free particle. In this sense these particles are abstractions. We deduce their existence from things we can see. For example, we can see the track a charged particle makes in a detector because these tracks are macroscopic manifestations. Only after a chain of theoretical arguments are we persuaded that what we are observing has been made by an invisible object.

The point I want to make is that this connection is not new. At the end of the nineteenth century there was a debate about the existence of atoms. No one doubted the chemist's atom, because its physical properties were largely irrelevant; it was merely a symbol that helped keep track of chemical reactions. But many physicists at the time actually insisted that the atom was real. Thermodynamic parameters such as heat, they said, were a manifestation of the motion of unobserved molecules. This accounted for the laws of thermodynamics. But there were skeptics, including the Austrian physicist-philosopher Ernst Mach. Mach argued that the laws of thermodynamics did not need an explanation, above all by invoking the actions of these abstract objects. To people who challenged him about the existence of atoms, he would say, "Have you seen one?" Nothing could change his mind. One wonders what he would have said about photographs that apparently show single atoms. Still, we have no photograph that shows a single neutron, let alone the more exotic particles in our palette. We accept their existence because it explains what we do see. As we proceed we will see how this works.

As I mentioned, the more recently discovered particles are more and more abstract. They are not needed to explain what goes on around us unless we happen to be in a high-energy physics laboratory. The discovery of their properties has become more and more expensive. My guess is that the cost of all the discoveries I will mention in Part I, when put together, was a few thousand dollars. I was the house theorist for the Harvard Cyclotron from 1955 to 1957. That machine, including the building, cost something like a half million dollars. The experiments now being done to find the Higgs boson cost billions. There is an irony in this, in that we now know that over 80 percent of the matter in the universe consists of objects we cannot identify. None of the particles I will describe is a candidate, so we have a mystery. I am personally pleased with this. I see nature as something like those nested Russian dolls, except in this case I think there may be no limit. The deeper we probe, the more will be left to probe.

Part I

Primary Colors

I
The Neutron

It has so far been assumed that the neutron is a complex particle consisting of a proton and an electron. This is the simplest assumption and it is supported by the evidence that the mass of the neutron is about 1.006, just a little less than the sum of the masses of a proton and an electron. Such a neutron would appear to be the first step in the combination of the elementary particles towards the formation of a nucleus. It is obvious that this neutron may help us to visualise the building up of more complex structures, but the discussion of these matters will not be pursued further for such speculations, though not idle, are not at the moment very fruitful. It is, of course, possible to suppose that the neutron may be an elementary particle. This view has little to recommend it at present, except the possibility of explaining the statistics of such nuclei as N^{14}.

—James Chadwick, "The Existence of a Neutron," *Proceedings of the Royal Society* A 136 (1932): 706

The word *isotope* was first used in the December 4, 1913, issue of the journal *Nature* in an article written by the British radiochemist Frederick Soddy, then a lecturer at the University of Glasgow. He had observed two versions of the element uranium that seemed to have the same chemical properties— they were both uranium—but slightly different masses. At a family lunch he brought up the problem of how to refer to these atomic brothers. A guest was a Scottish doctor named Margaret Todd, who was a friend of his in-laws. She came up with the name *isotope: topos* being the Greek for "place" and *iso* being the Greek for "same." So *isotope* means "same place"—the place being the position in the periodic table—and *isotope* it has been ever since. But we owe the physics of isotopes to the New Zealand–born physicist Ernest Rutherford, one of the greatest experimental physicists who ever lived.

By 1909 he had come back to England from Canada and was at Victoria University in Manchester. He had an assistant named Hans Geiger (later of the counter). Geiger told Rutherford that there was a student named Ernest Marsden who was looking for something to do, so Rutherford assigned both of them to a project. At the time, the nature of the atom was already being discussed. It was known that under normal circumstances it was electrically neutral and that it appeared to consist of a balance of negative and positive charges. The negative charges were known to be electrons—particles that had been discovered by the British physicist J. J. Thomson. The positive charges later came to be called protons (the term was invented by Rutherford), and they apparently carried most of the mass. In Thomson's model—known as the "plum pudding model"—the positive charges produced a kind of "pudding" of charge, while the much less massive electrons were scattered around like plums in that pudding. It was a consequence of this model that if a massive energetic particle acting as a projectile hit a thin metallic foil, it would pass through the foil like bullets through fog. This is what Rutherford expected to happen, but for reasons he could never explain he told Geiger and Marsden to keep an eye out for collisions in which the projectile scattered off the target at large angles. In the experiments that began in 1909 the target was a very thin gold foil and the projectiles were so-called alpha particles—later shown to be helium nuclei—produced in the decay of radium. Much to the scientists' astonishment, some of the alpha particles bounced off at very large angles. Some years later Rutherford reflected on this:

> It was quite the most incredible event that has ever happened to me in my life. It was almost as incredible as if you fired a 15-inch shell at a piece of tissue paper and it came back and hit you. On consideration, I realized that this scattering backward must be the result of a single collision, and when I made calculations I saw that it was impossible to get anything of that order of magnitude unless you took a system in which the greater part of the mass of the atom was concentrated in a minute nucleus. It was then that I had the idea of an atom with a minute massive center carrying a charge.

The calculations that Rutherford was referring to were published in 1911. The phenomenon they described has been known as Rutherford scattering ever since.

With the discovery by Soddy of isotopes, however, it was clear that there was a dilemma: mass did not follow charge. In his 1921 Nobel Prize lecture Soddy noted of isotopes, "Put colloquially, their atoms have identical outsides [the electrons] but different insides [the content of the nuclei]." The proton content of two isotopic nuclei was the same, but there had to be different electrically neutral components to account for the mass differences. It would have been considered extravagant at the time to invent a new particle for this purpose when you had perfectly good particles lying around—the proton and the electron—that you could stick together.

I now want to proceed in two steps. First I will explain how James Chadwick (see Figure 1) was able to discover this particle, and then I will explain why his dismissal of what was then known about the isotope nitrogen-14 was totally misguided. This will necessitate that I introduce you to the notion of spin.

Chadwick's experiment, done in 1932, is surprisingly simple to describe, although it required considerable skill to carry out. In fact, essentially the same experiment had been done previously by the Joliot-Curies in France, but they misinterpreted the results. They had studied certain emissions from an isotope of beryllium. These emissions are electrically neutral, and the Joliot-Curies assumed that they were very energetic quanta of electromagnetic radiation. These quanta were made to impinge on materials containing hydrogen, and very

Figure 1. James Chadwick. Photo by William G. Myers, courtesy of the American Institute of Physics, Emilio Segrè Visual Archives.

energetic protons were ejected after the collision. The Joliot-Curies thought this was a form of what was known as Compton scattering. The American physicist Arthur H. Compton, who had scattered radiation quanta from electrons, had been able to observe that these collisions obeyed the same kind of conservation laws of momentum and energy as did the elastic collisions of billiard balls. It was a good guess on the part of the Joliot-Curies that this was what was going on here—except that it was wrong.

Chadwick was suspicious. There were two things that bothered him. In 1928 the physicists Oskar

Klein and Yoshio Nishina produced a formula that described the details of the Compton results very well. Chadwick said the same formula should apply to the scattering of these quanta by protons, and the Joliot-Curie results did not fit. Also, he did not see where such energetic radiation quanta would come from. Hence he decided to repeat these experiments but this time use targets of a very general nature. In all cases he found high-energy protons ejected. He realized that a hypothesis that fitted everything was to suppose that this radiation consisted not of electromagnetic quanta but rather of neutral particles of about the same mass as the proton. Hence the neutron.

Now I will explain the evidence that Chadwick dismissed as to why these particles were elementary. This will take us into the concept of spin.

Spin was one of the triumphs of what became known as the "old" quantum theory. This is to be contrasted with the "new" quantum theory, which began in 1925 with the work of Werner Heisenberg. The guiding spirit of the old quantum theory, and indeed of the new, was Niels Bohr. After getting his Ph.D. in his native Denmark, he came to England to do postdoctoral work, initially with J. J. Thomson, but this did not quite pan out, so he went to work with Rutherford instead. Rutherford did not have a great deal of use for theoretical physicists, who he felt talked a good deal of moonshine, and in any event Rutherford himself was quite capable of producing whatever theory he needed. But an exception was Bohr. Rutherford saw in the gangling, rather awkward young man who had come to work with him the sparks of genius. The two developed a very close friendship that lasted the rest of Rutherford's life.

The nuclear atomic model posed a problem that seemed, and indeed was, beyond classical physics. Classical electrodynamics taught that an electric charge that accelerates radiates, removing energy. Now, the Rutherford model of the atom had all the electrons outside the nucleus, presumably accelerating all over the place. If so, why did they not lose energy and collapse into the nucleus? But this was only part of the problem. It was well known that when a gas of something such as hydrogen was excited, radiation was emitted in beautiful spectral patterns. How was this possible if the electrons were crashing chaotically into the nucleus? (An image that has always appealed to me is tossing a grand piano out of a window and expecting it to play Beethoven's *Moonlight Sonata* as it hits the ground.) This was the dilemma that Bohr set out to work on after he left England. In this he was helped by an unexpected source: a Swiss schoolmaster named Johann Jakob Balmer. Balmer, who was actually a mathematician, had noticed some twenty-five years earlier that if the wavelengths of the emitted light in the hydrogen spectrum were plotted, their values were related by a simple algebraic formula. This formula describes what is known as the Balmer series. Hence Bohr's problem was to find an explanation for the Balmer formula.

He employed what one can identify as three important ideas:

1. Not all orbits were allowed.
2. Radiation was emitted only when an electron jumped from some orbit to one characterized by a lower energy. Indeed, the energy emitted was equal to the difference of the energies associated with these orbits. Conversely, you could excite an electron by supplying this energy difference externally, which accounted for what is known as absorption spectra. The orbit of lowest energy was stable since there was no place for the electron to jump.

3. As the orbital energies got larger the orbits got closer and closer together in energy, so the classical situation was approached. Bohr called this the correspondence principle, and it played a role in his thinking from then on.

From these three principles Bohr was able to derive Balmer's formula for his series. It was an epoch-making discovery. What is remarkable to me is that it was done in the context of the old quantum theory despite the fact that in a fundamental way it did not make a lot of sense. For example, what were these jumps? Was the electron following some sort of trajectory when it made one? If so, what was it? The old quantum theory was mute on such matters.

Another great triumph, also done in the context of the old quantum theory, was the introduction of spin. The story of spin illustrates the fact that the path to such discoveries is often not a straight line but rather more like a random walk. In fact, as I will explain, the spin of the electron was actually measured in 1921, some years before the concept was even defined. It was only later that it was realized that what the experimenters Otto Stern and

Walther Gerlach had measured was indeed the electron spin. As the name might suggest, spin is some sort of angular momentum, but what sort? Though, as we shall see, spin is a quantum mechanical concept, here I will give a picture that is classical, for we are classical beings and that is how we tend to see the world. I have known a few mathematicians who claim that they can visualize four dimensions or more, but that is beyond most of us.

The Earth revolves around the Sun in an orbit that is nearly circular. This, along with the tilt in the Earth's axis, accounts for the seasons. There is a momentum associated with this motion that is called the orbital angular momentum. It depends on the speed of the Earth, its mass, and its distance from the Sun. But in addition the Earth is spinning. This rotation accounts for the night and day variation. The electrons in their Bohr orbits around nuclei have orbital angular momenta. Since the radii of these orbits are "quantized"—they have prescribed values—so do the orbital angular momenta. Bohr discovered that if he quantized the angular momenta, this would lead to the allowed orbits. If the angular momenta of the electrons are added up, this gives the atom an angular momentum. (There is also a contribution from the nucleus, which we can neglect here.)

What Stern, who conceived of the experiment, proposed to do was to measure the angular momentum of the silver atom. The idea was that because of the circulating electrons the atom acted like a tiny magnet. The strength of this magnet was proportional to the orbital angular momentum. If this tiny magnet was allowed to interact with a suitably designed magnetic field, the resulting force would cause the atoms to deviate from a straight-line trajectory. This deviation would depend on how the atomic magnets were oriented, which meant in what direction the angular momentum vector pointed. If the situation were classical, this direction would be arbitrary, so a smear of atoms would appear on the detector. On the other hand, if in fact the angular momentum vector was quantized, it could only point in a fixed number of directions. This meant that instead of a smear you would see lines that Stern called "space quantization." Bohr had an obscure argument that seemed to show that there should be only two such lines. This is what the experiment found, so there was general satisfaction. But only in 1927, thanks to the work of the Scottish physicist Ronald G. J. Fraser, would it be realized that the whole angular momentum of the atom had to do with the spin of one electron and that this measurement had ascertained its value. In the meantime, the notion of spin had to be invented.

This invention was triggered by a dilemma involving atomic spectra. Cadmium atoms give off a spectral line when excited; in 1896 the Dutch physicist Pieter Zeeman discovered that if cadmium atoms were put in a magnetic field, the one line would split into three, with lines above and below the original line. This could be explained classically, so it became known as the "normal" Zeeman effect. But by the 1920s lines began appearing that could not be explained classically. This became known as the "anomalous" Zeeman effect, although it is a more common occurrence than the normal Zeeman effect. These new lines could not be explained classically or by the old quantum theory. There were some desperate attempts using baroque atomic models, but finally in 1924 Wolfgang Pauli put his finger on the matter when he said in a paper that the effect was not one of the collective motions of the atomic electrons but must reside in the properties of the single electrons. This turned out to be right.

Pauli, whom I met a few times, was such a remarkable character that I cannot resist describing him. (He will reappear when we discuss the neutrino.) Robert Oppenheimer once said that Pauli, who was born in Vienna in 1900, was the only person he knew who was identical to his caricature. When I first encountered him in the 1950s he was a corpulent gentleman with a very large head. The Russian theoretical physicist Lev Landau invented a way of grading physicists. The worst was to have a small bottom and a small head; then you would have no good ideas and would not have the patience to work out any you did have. Next was large head and small bottom; then you would have good ideas but not the patience to work them out. He may have had Oppenheimer in mind here. The best was to have a large head and a large bottom; then you would have good ideas and would work them out. He must have had Pauli in mind. In fact, Pauli's head often bobbed rhythmically when he listened to a lecture. If it was a rapid motion, it meant that he disagreed with the speaker.

Things Pauli said and did have become legendary. One that I especially like was when he said that a paper he had read was not "even wrong." A young physicist was "so young and already so unknown." (I am convinced that this was a play on what the maid in *Die Fledermaus* said when she went to the masked ball and met a prince: "Still so young and already a prince?") Over the years he had many very gifted assistants, some of whom went on to win Nobel Prizes. One of these was Roy Glauber. When Glauber failed to write his mother often enough from Zurich, she complained to Pauli. Pauli never let Glauber forget this; it persisted for years. In the late 1950s Pauli came to Cambridge for a visit. A group of us went to the railroad station to greet him. It included Glauber and Victor "Vicki" Weisskopf, who also had been a Pauli assistant. Pauli greeted Glauber but did not mention his mother. He then went off with Weisskopf. The first thing he said to him was, "This time I fooled Glauber and did not say anything about his mother."

Pauli was a prodigy. He wrote an original paper about relativity just after graduating from high school. He decided that he would do his studies in physics with Arnold Sommerfeld in Munich. Sommerfeld was a legendary teacher, and among his students were several future Nobel Prize winners. Einstein came to lecture, and after the lecture Pauli commented, "Was Herr Einstein hat gesagt ist nicht so blöde"—"What Mr. Einstein has said is not so stupid." After Pauli took his Ph.D. in 1921, Sommerfeld suggested that he write a monograph on relativity. Pauli produced a masterpiece that is still one of the best sources on the subject. He seemed to have read every possible paper in every language. When he made his comment on the electron he was a lecturer at the University of Hamburg.

To me it is remarkable that Pauli did not see the implications of his own suggestion about the electron. Moreover, he vehemently opposed the correct explanation when it was first presented to him. Later he changed his mind and indeed provided the mathematics that we still use. The first correct suggestion was made by a very young German American physicist named Ralph Kronig. He had gotten a traveling fellowship from Columbia in 1925 and spent some of his time in Tübingen, Germany, where he was associated with a well-known spectroscopist named Alfred Landé. Landé showed him a communication from Pauli, who was about to visit. In it Pauli made the point about the effect residing in every electron singly.

It immediately struck Kronig that this requirement could be met if you gave the electron an extra degree of freedom that resembled an angular momentum. This was not an orbital angular momentum but an intrinsic property of the electron, which became known as spin. In the units that were being used, this spin was ½, which meant that in a magnetic field the electron could be either aligned or antialigned with respect to the field. In the jargon of the subject, the spin was either "up" or "down." This solved the problem of the anomalous Zeeman effect, but at an apparent cost. This was still the old quantum theory, which was an uneasy mixture of classical and quantum physics, and in it the electron was visualized as a ball of charge with a certain radius. Pauli realized that if you took what seemed to be a plausible radius, the surface of the sphere would need to be moving faster than the speed of light in order to provide the angular momentum. Moreover, there was a factor of two in the argument that was unaccounted for. On the basis of this he persuaded Kronig not to publish.

Meanwhile, unknown to everyone, there was a parallel enterprise afoot in Holland, at the university in Leiden, involving two very young Dutch physicists named Samuel Goudsmit and George Uhlenbeck. Goudsmit was still a graduate student, while Uhlenbeck was the assistant to Paul Ehrenfest, a senior professor and one of the best physicists in Europe. Ehrenfest suggested that Uhlenbeck act as a tutor to Goudsmit. Goudsmit knew all the spectroscopic data, while Uhlenbeck knew more theory. They hit on the idea of the spin but were too naive to see the apparent difficulties. They wrote a paper, which Ehrenfest submitted for them. Then they learned about the problems and tried to withdraw the paper, but Ehrenfest said it was too late, and besides, they were so young that even if it was a mistake, no one would notice. Indeed, very soon the difficulties evaporated: the electron began to be thought of as a point particle rather than as a ball of charge, and the factor of two was resolved. Thus in 1927 it was realized that the Stern-Gerlach experiment had actually measured the spin of the electron. The reason is that the silver atom has one valence electron outside the rest of the electrons. This electron determines the angular momentum of

the atom, since the rest of the electrons collectively—
the core—have no angular momentum. The fact
that Stern and Gerlach had gotten two spectral
lines meant that this angular momentum of this
valence electron was $\frac{1}{2}$.

Now after this detour we can return to Chadwick.
It will be recalled that Chadwick had dismissed the
evidence from nitrogen-14 that the neutron was an
elementary particle. What was this evidence? It was
then known that the angular momentum of this
nucleus was 1. It was also known that the spin of the
proton was $\frac{1}{2}$ and that one of the possibilities was
that there were seven protons and seven neutrons
in this nucleus. (The other possibility, which was
that there were fourteen protons and seven electrons, led to the same result.) But these seven protons must have an angular momentum that was in
total a half odd integer. If the neutron were a bound
electron and proton, it would have an integer spin,
and if you added up seven of these, you still got an
integer spin. There was no way you could get an angular momentum of 1 from these neutrons and protons. So it had to be that the neutron was an elementary particle of spin $\frac{1}{2}$. It did not take long for
everyone, including Chadwick, to accept this.

It was immediately realized that the discovery of the neutron opened up entirely new prospects for nuclear physics. Because it is electrically neutral, the neutron can probe deep inside the nucleus, since it is not repelled by the protons. One of the people who realized this was Enrico Fermi. In 1934 he assembled a small group in Rome that began bombarding various elements with neutrons, creating isotopes that had never been seen before. These experiments led to a discovery and a nondiscovery, both of which turned out to have great implications. At one point Fermi decided to put a piece of lead in front of his neutron target. He went to great trouble to get the lead properly shaped, but then for reasons he was never able to articulate he decided to replace the lead with paraffin. What happened was that the counters measuring the reaction rate jumped. This was before lunch, and so Fermi went home for a siesta; by the time he returned he had come up with the explanation. The neutrons had been slowed down by collisions with the hydrogen nuclei in the paraffin. But neutrons are both waves and particles, and this slowing down increased the wavelength and thus enhanced the nuclear reactions. (These so-called de Broglie waves have wavelengths that increase as the speed of the associated particle decreases.) This effect explains how moderators can be used to increase the nuclear reaction rates in reactors.

The nondiscovery had to do with uranium. I had a chance to discuss this with the physicist Emilio Segré, who was a witness. Working their way up the periodic table, the group eventually got to uranium, and they decided that they needed to put some extra shielding around the target. What they expected to find, and indeed what they claimed they had found, was the creation of new elements beyond uranium in the periodic table. In fact what they had found was nuclear fission, but the extra shielding prevented them from knowing this. What would it have meant for the world if nuclear fission had been discovered in Italy in 1934?

2
The Neutrino

On December 4, 1930, Wolfgang Pauli sent a letter to a group of colleagues who were attending a physics conference in Tübingen. He addressed them as "Dear Radioactive Ladies and Gentlemen." The letter was sent from Zurich, and Pauli apologized for not being able to attend the conference personally, as it conflicted with a ball he wanted to attend in that Swiss city. (In his day Pauli was a very good dancer and had a fondness for women. Figure 2, taken several decades later, shows that he had acquired a substantial avoirdupois.) The letter is one of the most remarkable documents in twentieth-century physics.

Pauli's concern was an anomaly that had occurred in experiments on what was known as beta decay. Ernest Rutherford, who had made an extended study of radioactivity, had identified three types of decay, which he called alpha, beta, and gamma. Gamma radiation had actually been discovered earlier by a chemist named Paul Villard.

Figure 2. Wolfgang Pauli (left) and Niels Bohr study the motion of a top. Photo by Erik Gustafson, courtesy of the American Institute of Physics, Emilio Segrè Visual Archives, Margrethe Bohr Collection.

Heavy nuclei such as that of plutonium can, in decaying, produce an alpha particle, which was identified as the nucleus of helium. Many other nuclei can decay to produce a gamma ray, which is a very energetic electromagnetic quantum. Some nuclei when they decay produce a beta particle, which is just an ordinary electron. It was in these decays that the anomaly manifested itself.

The obvious scenario for such a decay is for the parent nucleus to decay into a daughter nucleus and an electron. If energy and momentum are conserved in this decay and the parent nucleus is at rest, then the electron must emerge with one and only one energy. The parent nucleus has no momentum, which means that to conserve momentum the daughter nucleus and the electron must have equal and opposite momenta. This fixes the energy of the electron. The problem was that the experiment showed that the emerging electron had a spectrum of energies. This was such a puzzle that Bohr even proposed that energy was not conserved in the decay. Pauli thought that this idea was nonsense, and in his letter he made a counterproposal. He suggested that an invisible third particle was emitted with the other two and that this particle carried some of the energy and momentum. The particle was invisible since it was electrically neutral and interacted very weakly with everything. It simply departed from the scene of the decay.

I have no idea what the "radioactive ladies and gentlemen" made of this suggestion. Pauli was a very formidable physicist who had to be taken seriously. How seriously he took his own suggestion is unclear, since he never published it. But Enrico Fermi took it seriously and created the first real theory of beta decay. The question was what to call the particle. Pauli called it a neutron, but James Chadwick had already discovered the different particle that was soon widely being called the neutron. Fermi noted that *neutrone* means "big neutral one" in Italian, and since this particle, if it existed, had a small mass, he called it *neutrino,* "little neutral one"—and the name stuck.

The neutrino had an odd role in nuclear physics—sort of like a crazy uncle who was not quite all there. That was the attitude when I first learned about this particle in the early 1950s. It was almost a joke. This all changed in 1956 thanks to nuclear reactors, of which Fermi had created the first one during the war. These reactors are factories for producing radioactive fission fragments that in turn undergo beta decay. Hence an almost unbelievable flux of neutrinos is produced. In the Savannah River plant reactors in South Carolina a flux of over ten thousand billion neutrinos per square centimeter per second was produced in 1956 by the so-called P reactor. The two Los Alamos physicists who observed the neutrinos, Clyde Cowan and Fred Reines, in essence used a tank of water as the primary detector. When a neutrino was absorbed by a proton in the water, the proton was transformed into a neutron and emitted a positive electron. This process was the inverse of beta decay. (I shall discuss positive electrons in detail when I discuss antiparticles; suffice it to say here that when a positive electron meets a negative electron, both are annihilated, producing two very energetic gamma rays.) The neutron is absorbed by another nucleus, making the nucleus radioactive. It can emit a gamma ray, and indeed, when Cowan and Reines looked, they saw such gamma ray coincidences. But they found only about three such events per *hour*—a reflection of the weakness of the neutrino reactions. These experiments convinced everyone. Pauli was still around—he would not die until 1958—and one can imagine his feelings. Now, of course, neutrino experiments are commonplace and whole beams are produced at will.

The Cowan-Reines experiment was a small part of the 1956 "glorious revolution" in physics. It was a revolution that I have fond memories of, since I was there and even took a very small part. For readers to understand what was involved, I have to explain parity symmetry. In Figure 3 you will find two sets of axes that you can use to locate points in space.

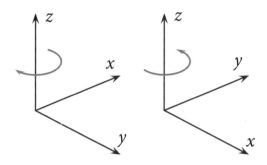

Figure 3. Parity symmetry. The left-handed system is on the left and the right-handed system is on the right.

Notice how the x and y axes are oriented. The one on the right is called right-handed because you can produce it naturally with two fingers and a thumb of your right hand. The one on the left is called left-handed because you can do the same with your left hand. Here is a challenge: Can you find a series of rotations that turns one of these coordinate systems into the other? It turns out that you cannot; you must invert one of the axes. This is called a parity transformation. The question is whether it matters from the point of view of the laws of physics. Until the "glorious revolution" the answer would have been an emphatic no. Indeed, some physics texts were written with one choice and some with the other. By the way, some popular descriptions of this call it "mirror symmetry." If you stand in front of a mirror with your right hand raised, your reflection will show this as your left hand. But if you think about it, it is because the light is reflected straight back at you, so I think it is better not to use this analogy.

Parity, like angular momentum, has two sources. Consider two like particles of spin 0 that orbit each other. In this state, the value of the parity depends on the angular momentum. For example, if there is no angular momentum, then by definition the parity is 1. The term of art is to say that the parity is "even." But there is a second source of parity that is intrinsic to the individual particles and which they carry around no matter what their state of motion is. This parity can be odd or even. The pions, of which we will hear much more later, have odd intrinsic parity. Any state of 2 has even intrinsic parity, while a state of 3 has odd intrinsic parity.

The "glorious revolution" of 1956 actually began in 1954. I was present at its inception, although I had no idea of what I was witnessing. I attended a conference of particle physicists that was held at the University of Rochester in January 1954. One of the

speakers was an Australian-born theoretical physicist named Richard Dalitz, who was then at Cornell University. Although we became good friends
later—he died in 2006—at the time I had never
heard of him. Dalitz spoke very fast in his Austra-
lian accent, and I did not understand the signifi-
cance of what he was saying; it took a couple of years
before the full implications became clear.

Dalitz was analyzing the decay of a then new
particle discovered in cosmic rays that had been
given the name tau. (It is now called the kaon, and I
shall have much to say about it later in the book.)
The tau decayed into three pi-mesons. Dalitz had
found a very clever way of plotting these decays,
which made it possible to read off the spin and parity of the tau. He concluded that the tau had spin 0
and odd parity. This was a useful thing to know but
hardly revolutionary, for by April 1956, at another
conference, things had taken a dramatic turn.

Dalitz had been able to add many new decay
events to his plot and had confirmed his original
result. But in the meantime a second particle had
been discovered and given the name theta. It had
nearly the same mass as the tau—indeed, within
experimental error the same mass—but it decayed

into two pions, which was a state of even parity. Hence on its face it looked as if this particle, if it was one particle, could decay into states of opposite parity. This was a real dilemma.

The visiting physicists shared rooms, and Richard Feynman was rooming with an experimental physicist named Martin Block. The theta-tau puzzle came up, and Block asked Feynman if the whole matter could be resolved if parity was not conserved in the decay. The next day Feynman reported Block's question to the conference (it is in the proceedings).

In the audience were two Chinese American physicists. Tsung Dao Lee and Chen Ning Yang. They were looking into a related question, whether existing experiments ruled out parity nonconservation in the kind of weak interactions that were responsible for these decays and others, but had not yet come to a conclusion. They then proposed an exhaustive list of experiments that might address the question directly. By June 1956 they had written a paper entitled "A Question of Parity in Weak Interactions." By the fall of the next year they had won the Nobel Prize for this work. I think that when they published their paper almost no one thought that parity would be violated. Pauli even bet money against it. But the experimenters set to work and in short order showed not only that parity was violated in these decays but that it was a very big effect. It is difficult to imagine the sensation this caused among physicists. At the time I was at Harvard as the house theorist for the cyclotron. but I was in close contact with the Physics Department and especially Julian Schwinger, who later won the Nobel Prize for his work in quantum electrodynamics. Schwinger, like almost everyone else, was sure that parity would be conserved. We gathered in his office to discuss the news. He began by saying, "Gentlemen, we must bow to nature." Many of us set to work to see what it meant. The following year I had a brief collaboration with Lee and Yang at the Institute for Advanced Study in Princeton. But what did these results say about the neutrino?

Until the "glorious revolution" the neutrino was pretty much thought of as Pauli had suggested it. The canonical example of a beta decay was the decay of the neutron into an electron, a proton, and a neutrino. (I ignore here the question of whether this is a neutrino or an antineutrino, something we will come back to when I discuss antiparticles later in the book. It is an open question at the moment whether the neutrino and antineutrino are really distinct particles, so for purposes of this discussion I will suppose that they are not.) The neutron has an average (mean) lifetime of about 881 seconds—under fifteen minutes. If you create a free neutron, this is how long you can expect it to survive. Pauli had suggested that, given the conservation of energy, the neutrino had to have a very small mass—less than the mass of the electron. The neutron is heavier than the proton by an amount that is a little greater than the electron mass, which is why it is unstable. If the neutrino had a larger mass than the electron, the beta decay would violate the conservation of energy since there would be more mass in the final state than in the initial one. But after the "glorious revolution" it was proposed that the neutrino emitted in beta decay had exactly zero mass, which meant that it moved with the speed of light. Experiments done at the time apparently showed that it was left-handed. Figure 4 makes the point.

In the diagram p stands for the momentum and S stands for the spin. The neutrino has spin $\frac{1}{2}$, and a left-handed neutrino has its spin and momentum oriented in opposite directions. This handedness of

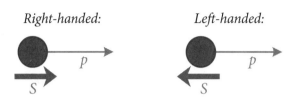

Right-handed: *Left-handed:*

Figure 4. Chirality.

the neutrino is called chirality—χειρ *(kheir)* being the Greek for "hand." With the massless neutrino it was shown that a theory could be invented in which this chiral property implied parity violation. Everyone was quite pleased with this connection until experiments several years later (which I will describe) showed that the neutrino does have a tiny mass. In physics, truth does not always equal beauty. There is no explanation for parity violation: it is a fact.

Not long afterward, a new puzzle suggested itself. The muon was much heavier than the electron. Hence the muon should decay into an electron and a gamma ray—$\mu \rightarrow e + \gamma$. But no such decay has ever been observed. There had to be a reason. The reason now accepted is that the muon and the electron are particles called leptons, which have a property associated with them that is called their lepton number. The muon and the electron have different lepton numbers, while the gamma's is 0, for it is not a lepton. Hence if lepton numbers are conserved, the gamma decay above is forbidden. But what about the neutrinos? To make things consistent, the neutrino emitted in beta decay is assigned the same lepton number as the electron. However, the pion, which has no lepton number, can decay into a muon plus a neutrino—$\pi \rightarrow \mu + \nu$. Hence this neutrino must have a muon lepton number, which makes it different from the neutrino emitted in beta decay. This is a testable proposition, and indeed it was tested in an experiment done in 1962 by Leon Lederman, Melvin Schwartz, and Jack Steinberger. They were awarded the Nobel Prize in 1988 for this work. What they did was to use very energetic neutrinos emitted in pi decay to produce muons, showing that this type of neutrino had the same lepton number as the muon. One may recall that when Cowan and Reines discovered the neutrino they used neutrinos to produce antielectrons, which shows that the neutrino emitted in beta decay has an electron lepton number. (We now know that there are actually three types of neutrinos, something I will come back to later.)

The Sun, like most stars, gets its energy from nuclear fusion. The process begins with a weak interaction that produces electron neutrinos. The physicist Ray Davis and the astrophysicist John Bahcall set up a detector—essentially a vat filled with cleaning fluid—deep inside the Homestake Gold Mine, located in Lead, South Dakota. They wanted it underground so as to filter out any particles except solar neutrinos, which pass undeterred through everything; in fact, a neutrino of a typical beta decay energy can pass through a light-year of lead without interacting. The reaction Davis and Bahcall were looking for was the solar neutrino transformation of the chlorine in their detector into an isotope of argon, which is possible if the solar neutrinos are of the electron type. They saw some reactions, but not enough—the theory predicted more. Bahcall tried to adjust models of the Sun, to no avail. A possible solution had been known to theorists for some time: if these two neutrinos had a different mass and if they moved with the same energy, they would move at different speeds, with the heavier one moving slower. Like all quantum mechanical particles, these neutrinos had a wave character. These waves could interfere, and over time one mass type could oscillate into another. What seems to have happened is that some of the electron neutrinos created in the Sun turned into muon neutrinos before they got to the Earth. The muon neutrinos could not produce the reaction Davis and Bahcall were looking for, which explains the result they saw. Unfortunately, neither this experiment nor its successors tell us the mass of the individual neutrinos; all they can give us is a relation among the masses. The relationship and the experiments show that any of the neutrinos must have masses at least a million times less than that of the electron. We are simply stuck with this fact, which has been confirmed with great accuracy in terrestrial experiments using accelerators. This means that the neutrino is not quite left-handed, as had been thought at the time of the "glorious revolution." Since the masses involved are very small, this effect was not seen in those early experiments.

John Updike, my erstwhile colleague at the *New Yorker*, was much taken by the neutrino and wrote an ode that begins, "Neutrinos, they are very small. / They have no charge and have no mass." The part about the charge is right, but they do have mass. In Appendix 3 I discuss some remarkable consequences of this fact. Updike claims that they "do not interact at all." If that were true, we never would have detected them. They interact some, though they do pass through the Earth and can, as he says, "enter at Nepal / and pierce the lover and his lass." I would be more concerned by the fact that there are about four microwave photons per cubic centimeter everywhere.

3
The Electron and the Photon

Some experiments in physics change paradigms. The experiments of 1956–1957 showing that in the weak interactions parity is not conserved are an example; physics was never the same again. This is also true, I think, of the experiments with cathode rays that J. J. Thomson described in an 1897 paper, which postulated the existence of what came to be known as electrons.

Thomson was born in 1856 in Manchester, England. His parents had decided on a career of engineering for him, but after his father died he went to Cambridge and never left. Ernest Rutherford was one of his students (see Figure 5). What strikes me about his experiments, and those of Rutherford as well, is that they were done on tabletops with a very small number of people involved. By contrast, the search for the Higgs boson, which I will describe toward the end of the book, involves detectors the size of a building and thousands of people.

Figure 5. J. J. Thomson (left) and Ernest Rutherford. Photo by
D. Schoenberg, courtesy of the American Institute of Physics,
Emilio Segrè Visual Archives, Bainbridge Collection.

Thomson's experiment involved vacuum tubes of the kind that I used to use as a boy when I tried to make radios. The tubes were glass bulbs with a metal filament inside them. When an electric current was applied, the filament glowed white hot and something was boiled off the metal that made an electric current in the tube. It was this current that was modulated to amplify the radio signal. I am embarrassed to say that, unlike Thomson, I did not have the slightest curiosity about what this current was; all I cared about was that the radio worked.

At one end of Thomson's tube was a cathode that was heated with electricity from a battery. This kind of experiment had been done before, and it was generally agreed that the particles that composed the discharge from the cathode had a negative charge. But the "ray" that traversed the tube was thought not to be composed of negatively charged particles, because earlier experiments had seemed to show that these particles could not be moved by applying an external electric field. Thomson felt that these experiments were flawed since the tubes had not been evacuated sufficiently. Once he created an adequate vacuum, he found that the ray's particles could indeed be moved by an electric field, and that they were negatively charged and presumably identical to whatever had boiled off the cathode. But how to measure their properties?

He used two methods, both involving electromagnetic fields. In the first method a magnetic field caused the cathode ray particles to move in a curved path. They accelerated and acquired kinetic energy, and this was transferred into heat when the particles collided with a solid body. Thomson measured the rise in temperature and hence could learn what the kinetic energy had been. This he related back to the curvature of the path, which in turn was related to the magnetic field. Putting these things together, he was able to get a value of the ratio of the mass of the particles to their charge. The second method he used, crossing electric and magnetic fields adjusted so that their effects just cancelled out and the particles passed through in a straight line, led to the same ratio, which turned out in all cases to be about a ten-millionth (in symbols, $m/e \sim 10^{-7}$ in suitable units). What to make of this result?

Thomson knew that there were experiments that measured m/e for ionized hydrogen. The atomic nucleus had not yet been discovered, so he did not know that ionized hydrogen was the proton. These experiments show that for this particle m/e was about 10^{-4}—a thousand times larger. He now had choices. He could say something about the m of his cathode ray particles, about their charge e, or about a mixture of the two. Thomson made the wrong guess. He guessed that the m/e of his "corpuscles," as he called them, was due to "the largeness of e as well as the smallness of m." He did make the right conclusion that they were a new state of matter—not an atom but probably a constituent of an atom. We now know, of course, that the electron has an equal although opposite charge to the proton and that its mass is closer to two thousand times smaller.

The particles we have discussed and will discuss appear to fall into two categories. There are those such as the neutrino that were predicted by theorists—Pauli in this case—and only turned up much later in experiments. And there were those such as the electron that had not been predicted but just turned up. The photon—the quantum of radiation—belongs more to the former category than to the latter. We can trace its genesis back to the year 1900, when the German physicist Max Planck presented some new ideas on what is known as blackbody or cavity radiation. Suppose you enclose a cavity with some walls. For the purposes of what we are going to discuss, neither the size nor shape of the cavity matters; neither does the material that the walls are made of. Now we heat the walls. The atoms absorb the heat and begin to oscillate. Accelerated charged particles radiate, so radiation is produced by the walls. This radiation bounces around the interior of the cavity, being absorbed and reemitted frequently by the electron oscillators. In the course of time an equilibrium is established and the walls and the radiation acquire the same temperature. Now one can ask, what is the distribution of this radiation? What I mean is that the radiation has a spectrum of frequencies. They are present with various intensities, and one wants to plot a curve of these intensities as a function of the frequency (or, alternatively, as a function of their wavelengths). What appealed to Planck was that this curve was universal. It was the same for blackbody radiation in equilibrium no matter what the cavity was. It depended only on the temperature.

There had been attempts to produce the functional form of this curve prior to Planck, notably by the German physicist Wilhelm Wien. These curves fitted the limited data that were available. But at the turn of the century Planck's Berlin colleagues had done new experiments that did not fit the Wien curve. Planck learned this at a Sunday lunch in his house, and that evening he set to work. By guesses and extrapolations he produced a new curve. To call what he did a "derivation" is gilding the lily. But his curve fitted the new data and has fitted every bit of data on blackbody radiation ever since.

On November 18, 1989, the Cosmic Background Explorer (COBE) satellite was launched. Its job was to measure the distribution of the radiation left over after the Big Bang. The universe is a large container, and the radiation produced should rapidly acquire a blackbody distribution. Some of this curve had already been measured, but COBE did the whole job. The curve that resulted from these data cannot be visually distinguished from Planck's theoretical curve.

Having produced his curve, Planck wanted to derive it from first principles. He found a derivation, but he did not like it at all. A classical oscillator can emit and absorb radiation on a continuum of frequencies. But to derive his curve Planck had

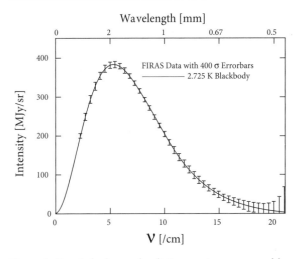

Figure 6. Cosmic background radiation spectrum measured by the Far Infrared Absolute Spectrometer (FIRAS) aboard the Cosmic Background Explorer satellite. The 2.725K refers to the present temperature of the radiation measured in degrees Kelvin. There are small departures from this curve that indicate turmoil in the very early universe.

to assume that his oscillators emitted and absorbed energy in packets, which he called quanta. The energy of a quantum, Planck argued, was proportional to its frequency, ν. The constant of proportionality Planck called h, so $E=h\nu$. Planck was very proud of this new universal constant h, and he told his son Erwin that it was one of the greatest discoveries in all of physics. But he did not like the idea of the quantized oscillators. My first great teacher in physics, Philipp Frank, from whom I heard about this as a college freshman at Harvard, said it was like the buying and selling of beer, which was done only in pints and quarts. Planck spent several years trying to derive his distribution without the use of quanta but failed. Enter Einstein.

The first of Einstein's great papers of 1905 has the title, translated into English, "Concerning a Heuristic Point of View about the Creation and Transformation of Light." *Heuristic* is such a strange word in the title of a scientific paper. It means something like "an aid to understanding" or "a trial that might lead to an understanding." I think that Einstein wanted to make clear that he was not offering a derivation. He understood clearly that Planck's formula could not be derived using the physics that was then known. Rather, he wanted to see what the formula meant.

The curve in Figure 6 has two limiting situations. For the low frequencies below the peak—which correspond to long wavelengths—the curve can be derived from classical physics. But above the peak—the high frequencies—it is fitted by the Wien distribution, and this cannot be derived from classical physics. Here is where the new physics was. Einstein decided to focus on this and to take the Wien distribution as an empirical fact. Prior to 1905, Einstein had mastered the kinetic theory. This is the theory that describes the average behavior of huge numbers of molecules, as in a gas. What Einstein showed was that as a consequence of the Wien distribution the radiation in the blackbody was behaving like a gas. Here is the way he put it: "According to the presently proposed assumption the energy in a beam of light emanating from a point source is not distributed continuously over larger and larger volumes of space but consists of a finite number of energy quanta, localized at points of space, which move without subdividing and which are absorbed and emitted only as units." Using Professor Frank's image, not only is beer bought and sold in pints and quarts, but wherever you find it, it is only in pints and quarts.

What I find remarkable in Einstein's paper is that nowhere does he say that these quanta are particles. He must have known that a quantum has an energy $h\nu$ and a momentum $h\nu/c$ where c is the speed of light. Why doesn't he say it? And why does he not say that to be consistent with the relativity theory these particles have to be massless? All particles that are massless move with the speed of light and vice versa. He must have known. In any event, he used the conservation of energy to explain some results in what is known as the photoelectric effect. Light is shined on a metallic surface and electrons are emitted. The number of electrons emitted depends on the intensity of the light, but the energy of the electrons depends only on the frequency of the light, according to Planck's law. It was for this work that Einstein was awarded the Nobel Prize in physics for 1921; the Royal Swedish Academy of Sciences wanted nothing to do with relativity and said so. (The prize money went to Einstein's first wife as part of their divorce settlement.)

In 1923 Arthur Compton scattered these quanta from the electrons in the carbon atom. If you think of the collision as obeying the same sort of energy and momentum conservation laws as the elastic scattering of two billiard balls, then Compton's results could be accounted for. This demonstrated that the quanta were particles. The name for this type of particle comes from an unexpected source—the chemist Gilbert N. Lewis. In 1926 he published a letter to *Nature* entitled "The Conservation of Photons." It contains the sentence "I therefore take the liberty of proposing for this hypothetical new atom, which is not light but plays an essential part in every process of radiation, the name *photon*." The rest of the paper is pretty confused: of course photons *are* light, and they are *not* conserved. The only thing that remains from this paper is the name for the particle. The theory that treats only photons and electrons is known as quantum electrodynamics, and I will come back to it later. But these are the particles physicists dealt with until the early 1930s—the primary colors. In the next chapter we will venture further into the palette.

Part II

Secondary Colors

Figure 7. Hideki Yukawa.
Photo courtesy of Shuki Zakai.

4

The Pion and the Muon

By the mid-1930s the number of elementary particles that had actually been observed could be
counted on the fingers of one hand. There were the
proton, the neutron, the photon, the electron, and
the positron. The last is the antiparticle to the electron. (I will later devote a chapter to antiparticles,
but here I can note that the positron has the same
mass as the electron and the opposite charge. Electrons and positrons can annihilate into a pair of
gamma rays, which are very high-energy photons.)
By the time World War II broke out, the muon had

been added to the list. We begin with the pion, which was not observed until 1947.

We have seen that in 1909 Rutherford along with his two young colleagues discovered the atomic nucleus, in which most of the mass of the atom was concentrated. It is tens of thousands of times smaller in diameter than the average distance to the electrons. The chemical business of the atoms is done by the electrons with the nucleus as a by-stander. This shows incidentally that the force of attraction between the electrons and the protons in the nucleus must act over a long range, well beyond the nucleus itself. But the protons repel each other electrically. If there were not a counteracting force, the nucleus would come apart. This counter-acting force must be stronger than the electric force and must be very short-range. This was the puzzle. Enter Hideki Yukawa.

Yukawa was born in Tokyo in 1907. His father became a professor of geology at Kyoto Imperial University, and Yukawa did his studies there. He took his Ph.D. in 1929, and when the Depression began in Japan and he could find no job, he became an unpaid assistant to a professor of theoretical physics while continuing to live with his family. It is interesting that he was able to keep up via journals with the rapidly developing quantum theory. In 1932, after the neutron had been discovered, Heisenberg wrote three papers on nuclear forces. Like Chadwick, he believed that the neutron was a bound electron and proton, which meant that there would be electrons in the nucleus—something that led to all kinds of difficulties, which vanished the minute one accepted the fact that the neutron was indeed an elementary particle. Yukawa used this idea to create his own theory of nuclear forces.

The clearest way to describe what he did is to use Feynman diagrams (although this is anachronistic, since Feynman diagrams were not invented until after the war). I begin with the force that causes two electrons to interact. In the quantum electrodynamic picture this force is caused by the exchange of photons (see Figure 8).

Here two electrons are shown exchanging a virtual photon. It is "virtual" in the sense that this photon is not directly observable. One can work out the mathematics associated with this diagram and show that in the first approximation it leads to a force between the two electrons that is the same as the one French physicist Charles-Augustin de

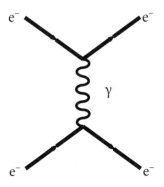

Figure 8. Feynman diagram.

Coulomb at the end of the eighteenth century claimed acted between charged bodies. In this case the force falls off in strength as the square of the distance between the two charged bodies. One can show that this is related to the fact that the particle being exchanged—the photon—has zero mass. For many reasons this force cannot be what holds the nucleus together: for one, it is too weak, and for another, it is why protons repel each other.

Yukawa, as Figure 9 shows, replaced the electrons by what is called in the diagram a pi-minus particle, though Yukawa called it a "heavy quantum."

He was free to take the coupling of the heavy quantum to the neutrons and protons at any strength

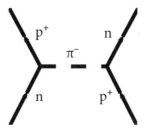

Figure 9. Feynman diagram showing pion exchange.

he liked, so he could choose it to overpower the electric force. Moreover, he could choose any mass he wanted for it, but he had a clue from quantum mechanics. The neutron and proton have about the same mass, while the pion also has mass. Hence the emission and absorption of this massive pion violate the conservation of energy. One of Heisenberg's uncertainty principles tells us that the conservation of energy can be violated provided that we don't do it for too long. The shorter the time of observation, the more uncertain is the energy. Suppose we call the violation ΔE and the time interval Δt. Then the uncertainty relation tells us that $\Delta E \Delta t \sim h$, where h is Planck's constant. This constant is what measures our intrusion into the quantum world. Using Einstein's relation between mass and energy, $\Delta E = mc^2$, where c is the speed of light, if we assume that our virtual particle is moving with the speed of light, then the distance it can travel is about $\Delta t c$. This is the range of the force. Putting things together, we have $\Delta t c \sim h/mc$. Thus we see the relationship between the range of the force and the mass of the exchanged particle. But Yukawa argued that the range of the nuclear force must be about the size of the nucleus, of which he had some idea. This enabled him to guess that the mass of the heavy quantum had to be about two hundred times that of the electron. In 1934 he wrote a paper in excellent English that was published in a distinguished Japanese journal the next year and then was totally ignored. But in 1936 a particle was discovered in cosmic rays that had about the mass Yukawa had predicted. This called attention to Yukawa's paper. However, it soon became clear that this particle, which first became known as the mu-meson and is now called the muon, had no strong interactions. It was for all practical purposes a heavy electron. (When the physicist I. I. Rabi heard about it, he famously remarked, "Who ordered *that*?") The muon was unstable and decayed into an electron and two neutrinos. It had nothing to do with the nuclear force.

The real Yukawa particle was found in cosmic rays in 1947. It became known as the pi-meson (later the pion) and came in three varieties, with positive, negative, and zero charge, respectively. It had about the mass that Yukawa had predicted. The charged varieties decayed rapidly into muons and neutrinos, which is why the muon was the first to have been observed: the parent pi-mesons had decayed away. The neutral pion decays into two very energetic gamma rays. For this work Yukawa was awarded the 1949 Nobel Prize—the first Japanese to have earned one. A sense of what this meant to national pride coming so soon after the war can be seen in the postage stamp reproduced in Figure 7. It shows Yukawa in formal attire.

This was all very well except that the same sort of methods that were used to detect the pi revealed a completely unexpected plethora of new mesons. To paraphrase Rabi, who ordered *them?* Something of the feeling is nicely summed up in a bit of doggerel written by none other than Edward Teller, who is not generally known for his light verse. The "Blackett" he refers to is Patrick Blackett, a British physicist who won the Nobel Prize for his work in particle detection.

There are mesons pi, and there are mesons mu.
The former ones serve us as nuclear glue.
There are mesons tau—or so we suspect—
And many more mesons which we can't yet detect.

Can't you see them at all?
Well, hardly at all,
For their lifetimes are short
And their ranges are small.

The mass may be small, and the mass may be large.
We may find a positive or negative charge.
And some mesons will never show on a plate,
For their charge is zero, though their mass is quite
   great.

What, no charge at all?
No, no charge at all.
Or if Blackett is right,
It's exceedingly small.

When I was doing my Ph.D. work at Harvard in the early 1950s, most of the people who were working with Schwinger and his associates were doing calculations in quantum electrodynamics, but a few, including myself, worked on pi-meson theory. My advisor, Abe Klein, favored a particular theory we were exploring; in reading the proceedings of the

1954 Rochester conference, I find mention of Klein and several others discussing this theory, as well as sessions devoted to new experimental results on pions. As I noted in Chapter 2, there was also Richard Dalitz discussing his method of plotting the decay of another newly discovered particle. But these new particles occupied a very small part of the program, which now looks antique—none of the matters that were discussed then are of the slightest interest from the present point of view. The model that Klein and I used was basically useless, except that it allowed me to get my degree. When I hear discussions nowadays of a "final theory," I think of these 1954 conference proceedings and wonder.

5
The Antiparticle

In 1928 Paul Dirac produced an equation for the electron that united the quantum theory with Einstein's relativity. Some people think that it is the most beautiful equation in theoretical physics:

$$\left(i\hbar\gamma^{\mu}\frac{\partial}{\partial x^{\mu}} - mc \right)\Psi(x^{\mu}) = 0$$

I am not sure I would go that far, but it is a very beautiful equation.

However, Dirac recognized at once that there were problems with its solutions. For each value of the momentum of the electron there were four solutions, and only two of them made sense. The other two gave the electron a negative energy that was impossible. This led to a couple of years of what Pauli called "desperation physics" before a solution was arrived at. If the electron had a negative charge and was represented by two of the positive energy solutions, then the other two solutions represented a particle identical to the electron but with the opposite charge. This was the introduction into physics of antimatter. The problem was that at the time this solution was arrived at, no such antielectron was known to exist. The matter was resolved when the Caltech physicist Carl Anderson found one in cosmic rays in 1932. It became known as the positron.

Once the positron was found, the whole subject of antimatter began to be taken seriously. It was realized that every particle had an antiparticle. In a few cases, such as that of the photon and the electrically neutral pi-meson, the particles and antiparticles are identical. Neutrinos have no charge, but it is an open question as to whether they are identical to their antiparticles. But if a particle had a charge e, then the antiparticle would have a charge $-e$. If a particle had a mass m, then the antiparticle would also have a mass m. In particular, the proton had a mass about two thousand times the mass of the electron and a charge of e. Thus there had to be an antiproton with the same mass and the opposite charge. How to find it?

Cosmic rays were not promising, since the antiproton annihilates with ordinary matter and what you see is the detritus. What had to happen is that an accelerator had to be constructed that was sufficiently powerful to produce the antiproton. The idea was to have a target such as liquid hydrogen, whose nucleus is just one proton, and then to produce an energetic beam of protons in the accelerator that would collide with these target protons. Out of this collision an antiproton would be produced, or so it was hoped. The question was how much energy this machine would need to have. Here there was a complication. One began with two protons—one in the beam and one in the target. Combined, these had two units of positive charge. The aim was to end up with an antiproton, with one unit of negative charge. To balance the charges, the final state would need to have three protons and one antiproton. The reaction thus would be $p + p \rightarrow p + p + p + p^c$, where p is the proton and p^c is the antiproton.

There is no simpler reaction. But what is the minimum proton beam energy needed? A straightforward calculation shows that this energy is about six times the mass of the proton. The problem was that at the time this calculation was made there was no accelerator in the world that had

Figure 10. This picture, taken in 1955 in Berkeley, California, shows the Bevatron under the dome. Photo courtesy of the Lawrence Berkeley National Laboratory.

beams of this energy. The only thing to do was to build one.

Planning for such an accelerator, which became known as the Bevatron—a reference to the energy it was to produce, which was measured in billions of electron volts—began in the late 1940s at Berkeley, California (see Figure 10). That it was in Berkeley was no accident. Ernest Lawrence, who was the inventor of the cyclotron, was a professor there. Not only was Lawrence an excellent physicist, but he was a genius at raising the kind of money needed to build such a machine. Originally the accelerator was designed to produce proton energies well above the threshold for creating antiprotons, but when the machine went into operation in 1954, the energy had been reduced, so there was only just enough to exceed the threshold. In 1955 the team of experimenters announced that the antiproton had been found. I don't think this came as a surprise to anyone in physics, but it seemed to open up a good deal of popular fantasy about antimatter. This was neatly summarized by a poem in the *New Yorker* that was signed only by the initials H. P. F. I knew that the initials were those of Harold P. Furth, who was a physicist and a classmate of mine from Harvard, where he wrote for the *Lampoon*. The poem:

The Perils of Modern Living

Well up above the tropostrata
There is a region stark and stellar
Where, on a streak of anti-matter
Lived Dr. Edward Anti-Teller.
Remote from Fusion's origin,
He lived unguessed and unawares
With all his antikith and kin,
And kept macassars on his chairs.
One morning, idling by the sea,

He spied a tin of monstrous girth
That bore three letters: A. E. C.
Out stepped a visitor from Earth.
Then, shouting gladly o'er the sands,
Met two who in their alien ways
Were like as lentils. Their right hands
Clasped, and the rest was gamma rays.

I do not know where the tropostrata are, but I do know where you can find a profusion of antiprotons—the inner Van Allen Belt. The Van Allen Belts are regions above the Earth where charged particles are trapped in the Earth's magnetic field. The inner belt extends from about sixty to about six thousand miles above the Earth's surface. When very high-energy protons get trapped there, the result can be antiprotons, produced using the same reaction as in the Bevatron. (There are not enough of these antiprotons to constitute a menace to space travel.)

A final word about antiparticles that are electrically neutral. First of all, as I noted earlier in the chapter, there is a class in which particle and antiparticle are identical. These include the neutral pi-meson and the photon. Then there is a class in which they are not. These include the neutron and the K-meson (the latter will be discussed in Chapter 6). The neutron and antineutron have many properties in common. Both are electrically neutral and have the same mass. Both have the same spin, ½. Both are unstable, but their decay mode is different. The antineutron decays into an antiproton, a positron, and a neutrino. This decay mode is dictated by something that is called the conservation of baryon number. The baryon number, which is attached to particles such as the proton and neutron, is 1 for these particles and −1 for the antiparticles. The rule is that this number is conserved in all processes. That is why the antineutron decays into an antiproton. If you go back and look at the reaction that created antiprotons, you will see that it reads $p + p \rightarrow p + p + p + p^c$, which conserves the baryon number. The free antineutron, apart from being unstable, annihilates with neutrons (and protons, for that matter), so you will not find free neutrons floating around.

But in 1965 a wonderful object was produced in a high-energy accelerator. We have seen that the nucleus of ordinary hydrogen consists of one proton. But there is a stable isotope called deuterium, whose nucleus (called a deuteron) consists of one proton and one neutron. In 1965 antideuterons were produced that consist of one antineutron and one antiproton. They have the same mass as deuterons.

Neutrinos may or may not have distinct antiparticles, as I remarked earlier in this chapter. Very difficult experiments have been under way for years trying to decide this, and the jury is still out.

6
Strange Particles

In prior chapters I have noted that some particles were discovered in cosmic rays, the positron being an example. Someone unfamiliar with the subject might get the idea that there was a kind of backyard treasure hunt in which these particles were unearthed. Since the particles I will discuss in this chapter were also found initially in cosmic rays, I want to explain what this means, beginning with a discussion of what a cosmic ray is.

In 1896 the French physicist Henri Becquerel made the accidental discovery that a substance containing uranium emitted charged particles. This was the discovery of radioactivity. Radioactivity was thought then to be the solution to a puzzle. Atmospheric air appeared to be ionized: it carried an electric charge. The assumption was made that this was caused by the natural radioactivity coming from the Earth. This was tested when in 1912 the Austrian physicist Victor Hess flew in a balloon to an altitude of some 5,300 meters carrying an electrometer. He found that the ionization quadrupled at this altitude, which meant that its source was extraterrestrial. At first it was thought that it was emanating from the Sun. But Hess ruled this out when he flew in his balloon during a nearly total solar eclipse and showed that the radiation persisted. But what was it and where did it come from?

In those presatellite days one could only study the radiation fairly close to the Earth. There were two schools of thought. One argued that the primary radiation consisted of very high-energy photons—gamma rays—and the other argued that it was positively charged particles. The two proposals could be assessed by measuring the cosmic ray flux at different locations on the Earth. Uncharged particles such as gamma rays would not be deflected by the Earth's magnetic field, while charged particles would. Furthermore, because of the Earth's magnetic field, it was predicted that more positively charged cosmic rays would come from the west than from the east. By the end of the 1930s measurements made it clear that the charged-particle people were right. Now it is agreed that most of the primary radiation consists of high-energy protons. Some of them are of a higher energy than can be produced in any accelerator. It is also generally agreed that they have their origin in supernova explosions. The ones that we see have been traveling for millennia in the vacuum of outer space. When they crash into our atmosphere they produce a great variety of secondary particles, which are what is detected. But how to detect them?

The detector of choice was the so-called cloud chamber, invented by the Scottish physicist Charles Wilson. Cloud chambers now come in a variety of forms, and one can find on the Web instructions as to how to make one for less than a hundred dollars. Wilson's consisted of a sealed chamber in which there was saturated water vapor—water vapor that was close to condensation. There was a diaphragm that could be expanded, thus cooling the mixture and causing the water vapor to condense. If a charged particle passed through, it would ionize a bit of the vapor. What one would observe was a trail of ions produced by these particles. To an untrained observer the whole thing could look like a Mondrian abstract. A fairly simple example is shown in Figure 11. It is one of Anderson's first cloud chamber photographs exhibiting a positron.

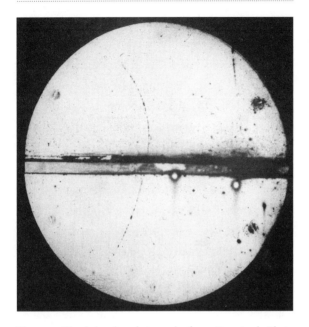

Figure 11. Cloud chamber photograph of a positron track. Photo by C. D. Anderson, courtesy of the American Institute of Physics, Emilio Segrè Visual Archives.

On the left you see a thin curved line. Anderson had a magnetic field acting on his chamber, and he was very familiar with how an electron curved in this field: as the particle traveled upward it lost energy and the curvature of the track increased. This track was identical except that it was curved in the opposite direction, meaning that the particle had a positive charge. It is interesting that in 1936 Anderson and his collaborator Seth Neddermeyer found other cloud chamber photographs that looked something like this. But the tracks were less curved, indicating that the particle producing them had a larger mass than the electron but less than the proton. Anderson called it a mesotron, to indicate that it had a mass in the middle between the proton and the electron. The name mesotron morphed into the name mu-meson; now, since it is clearly understood that it is a lepton and not a meson, it is known as the muon.

But where was Yukawa's particle? After the war a solution to the puzzle was suggested that turned out to be correct. It was hypothesized that cosmic ray protons produced Yukawa's particle, which rapidly decayed into a muon and a neutrino. The muon was the particle Anderson and others had found. The real pi-meson, or pion, was found in 1947 by a collaboration that used a photographic emulsion as a detector. This is a light-sensitive gelatin, so in some sense the charged particle takes a picture of itself. The pion is somewhat heavier than the muon, which is why it can decay into it. The charged pi-meson decays into a lepton and a neutrino with a lifetime of about a hundredth of a microsecond, while the neutral pi-meson decays into two gamma rays with a lifetime about a billion times shorter. This shows that two different kinds of forces are involved in these decays. The weak force produces the charged pion decays, while the much stronger electromagnetic force produces the more rapid neutral pion decay. In our exposition so far we have been dealing with three kinds of forces. There is the strong force, which holds the nucleus together. There is the electromagnetic force, which tries to push the protons apart, and there is the weak force, which produces decays into leptons. (There is also the force of gravitation—the weakest of all—which I shall discuss toward the end of the book.) If we compare the electric force between two electrons, which is repulsive, and the gravitational force, which is attractive, the electric force is stronger by a factor of 10^{36}.

The cloud chamber photo in Figure 11 is simple compared to the sort of immensely complex tangle of tracks that experimenters have to decode. Figure 12 is a famous example. It is a picture taken at Brookhaven National Laboratory in Upton, Long

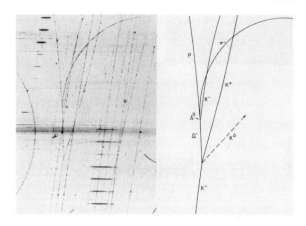

Figure 12. The discovery of the Ω^- at Brookhaven National Laboratory. Photo courtesy of the American Institute of Physics, Emilio Segrè Visual Archives.

Island, in 1963, and it shows the discovery of a particle named the omega minus (Ω^-) which confirmed the quark model of the elementary particles (which I shall discuss later). Note how it has been decoded with the labeled particles.

The Brookhaven picture was taken using a so-called bubble chamber. In it charged particles leave tracks, and it is up to the experimenter to figure out what they mean. One of the first examples of this was published in the journal *Nature* in 1947 by the British physicists G. D. Rochester and C. C. Butler.

They used a photographic emulsion. I have looked at this picture many times and I am still astounded by what they could deduce from it. Sometimes even if you know how a magic trick is done you are still amazed by it. They took 5,000 photographs during 1,500 operating hours of their cloud chamber. They saw two tracks that came together to make a fork or vee. What is significant is that the two tracks emerged out of nowhere. The experimenters found no other photographs of this kind. They decided that whatever was causing this had nothing to do with collisions. The only explanation they could think of was that it must be an electrically neutral particle that left no tracks but decayed spontaneously into two charged particles, which did leave tracks. They were able to estimate the mass of this object and came to the conclusion that, based on this data, it could not be smaller than about 800 times the mass of the electron (although with a very considerable error). They had a second photograph showing the decay of a charged particle that seemed to have a mass of about 1,000 times the mass of the electron (again with a substantial error). From the tracks it appeared that the decay products were pi-mesons, which in turn decayed into mu-mesons and neutrinos.

The parent particles were initially named V particles because of the tracks. For a while no further examples were found, but then in 1950 Caltech observers found fifty charged V particles and four neutral ones. By the 1950s it was understood that there were four types of these particles, which today we call K-mesons. There was a positive one, with a mass of about a thousand electron masses, and its antiparticle, with the same mass and the opposite charge. Then there were two neutral ones— one the antiparticle of the other—with a slightly larger mass than their charged counterparts. The neutral ones are truly remarkable, and we shall return to them.

In 1948 Brookhaven National Laboratory got the go-ahead to build the largest particle accelerator up to that time. It went into operation in January 1953. Figure 13 shows the ring around which protons were accelerated. Comparatively, the Cosmotron is almost a toy—it is much smaller than a single detector at the Large Hadron Collider—but it did make beams of K-mesons, which rendered the cloud chamber passé. I have a special fondness for

Figure 13. The Cosmotron. Note the human figures, which set the scale. Image courtesy of Pearson Scott Foresman.

this machine, for I was on the theoretical staff at Brookhaven for a couple of years in the 1960s. When the machine was down I used to go into the building at night to practice my trumpet. The acoustics were wonderful.

While some amazing discoveries were made using cosmic rays, there were limitations. Not only could one not get precise masses, but the reactions that led to the creation of the new particles were not evident. Moreover, there was no continuous supply of these particles: they arrived at the whim of the cosmic rays. This was all rectified once the Cosmotron went into operation and mesons were produced at will. Furthermore, a very striking thing was discovered about the production reactions.

Recall that the Cosmotron produced a very high-energy beam of protons. Suppose you had as a target liquid hydrogen. Then you would be studying proton-proton collisions. Also recall that each proton has a baryon number of 1, so initially in this reaction, which involved two protons, you would have a baryon number of 2. This number must be conserved in the production process. Let us write a reaction that produces a K^+ in the form $p+p \rightarrow K^+ + p + X^0$. To conserve charge and baryon number the X^0 must be a particle with no charge and baryon number 1. The obvious particle that comes to mind is the neutron. But the experiments showed that this reaction never happens this way—a neutron is never produced. So the X^0 must be a different particle altogether. In fact, such objects had already shown up in cosmic rays.

The lightest of these was the so-called Λ^0. It was a neutral particle about half again as massive as the proton. It decayed rapidly into either a π^- and a proton or a π^0 and a neutron. The other candidate was the so-called Σ^0. It was somewhat more massive than the Λ^0 and indeed decayed into it along with a gamma ray. The K-meson was always produced with one or the other of these. This was codified in the rule of associated production, but the rule didn't offer any explanation.

The next step was taken by Murray Gell-Mann, of whom we shall hear a great deal, and the Japanese physicist Kazuhiko Nishijima. I will present Gell-Mann's version, since it was the one that everyone used. Gell-Mann always had a way with names, and since the new particles were strange, he introduced degrees of "strangeness." Electrons, muons, neutrinos, and photons were excluded from the game. Pi-mesons, neutrons, and protons were assigned a strangeness of 0. The situation with the K-mesons was more complicated. The positive and neutral K were assigned a strangeness of 1, while their antiparticles had a strangeness of -1. The Λ^0 and the Σ^0 were assigned a strangeness of -1. Another, even heavier particle, the Ξ^0, had been discovered, and it had a strangeness of -2. It decays into a Λ^0 and a π^0, thus the rule of associated production was simply the statement that in the strong production interactions strangeness was conserved. This was not true of the interactions that caused the particles to be unstable. We have seen, for example, that K-mesons, which are strange, can decay into pions, which are not.

Not long after he had invented strangeness Gell-Mann went to the University of Chicago to give a lecture on it. In the audience was Enrico Fermi, who was then a professor at the university. During

Table 1. Mesons.

Particle	Symbol	Antiparticle	Makeup
Pion	π^+	π^-	ud^c
Pion	π^0	Self	$(uu^c - dd^c)/\sqrt{2}$
Kaon	K^+	K^-	$us^c, u^c s$
Kaon	K^0	NA	$ds^c,$
Antikaon	K^{0c}	NA	$d^c s$
Eta	η^0	Self	$uu^c + dd^c - 2ss^c$
Eta prime	$\eta^{0'}$	Self	$uu^c + dd^c - 2ss^c$
Rho	ρ^+	ρ^-	ud^c
Rho	ρ^0	Self	uu^c, dd^c
Omega	ω^0	Self	uu^c, dd^c
Phi	ϕ	Self	ss^c
D	D^+	D^-	cd^c
D	D^0	D^{0c}	cu^c
D	D^+_s	D^-_s	cs^c
J/Psi	J/Ψ	Self	cc^c
Upsilon	Υ	Self	bb^c

Note: The neutral kaons require special treatment to be explained. A superscript "c" stands for the antiparticle. "Self" means that the particle and the antiparticle are identical.

the course of the lecture Fermi asked a question. As a rule, he noted, there are no processes that can turn a particle into its antiparticle. For baryons this would be a violation of the conservation of baryon

Rest Mass (MeV/c²)	Strangeness	Lifetime	Decay Modes
139.6	0	2.60×10^{-8}	$\mu^+ \nu_\mu$
135.0	0	0.84×10^{-16}	2γ
493.7	−1/+1	1.283×10^{-8}	$\mu^+ \nu \mu, \pi^+ \pi^0$
497.6	+1	NA	NA
497.6	−1	NA	NA
548.8	0	5.0×10^{-10}	$2\gamma, 3\mu$
958	0	3.2×10^{-21}	$\pi^+ \pi^- \eta$
770	0	0.4×10^{-23}	$\pi^+ \pi^0$
770	0	0.4×10^{-23}	$\pi^+ \pi^-$
782	0	0.8×10^{-22}	$\pi^+ \pi^- \pi^0$
1020	0	20×10^{-23}	$K^+ K^-, K^0 K^{0c}$
1869.4	0	10.6×10^{-13}	$K + \_ , e + \_$
1864.6	0	4.2×10^{-13}	$[K, \mu, e] + \_$
1969	+1	4.7×10^{-13}	$K + \_$
3096.9	0	0.72×10^{-20}	$e^+ e^-, \mu^+ \mu^-, \ldots$
9460.4	0	1.21×10^{-20}	$e^+ e^-, \mu^+ \mu^-, \ldots$

number, for example. But the neutral K-mesons are quite different. Both the particle and antiparticle can decay into two pions. They have opposite strangenesses, but strangeness is not conserved in the decay. Why, Fermi asked, could not the neutral K and its antiparticle simply transform themselves back and forth, and if that was possible, what were the implications?

Gell-Mann returned to the Institute for Advanced Study in Princeton, where he was a visitor. He had been collaborating with Abraham Pais, who was a professor there and who had originated the associated production rule. Here is what they now came up with. Because of the associated production rule, the K^0 that is produced has a definite strangeness—1 in this case. But now it begins to transform into its antiparticle, so at any later time it is a mixture of particle and antiparticle, which have opposite strangeness. It is these mixed states that have definite masses and decay. There are two of these, one with a fairly short life that decays into two pions and one with a longer life that decays into three. Over the course of time the short-lived one disappears, leaving only the long-lived one. This leads to a delightful property. If this long-lived part is now made to interact with matter, the two strangeness components out of which it is made act differently. The effect of this is to revive some of the short-lived part, so the two-pion decay is once again observed.

By now the plethora of particles may be making you somewhat dizzy, even though so far I have let you off easy. Table 1 is a recent table of mesons.

Clearly matters were getting out of control. But things were about to look up.

Figure 14. Quark cheese. Courtesy of Appel Farms.

7
The Quark

> Three quarks for Muster Mark!
> Sure he hasn't got much of a bark.
> And sure any he has it's all beside the mark.
>
> —James Joyce, *Finnegans Wake*

In March 1963 Murray Gell-Mann came to Columbia University to give a lecture on a scheme he had recently invented for classifying elementary particles and which he had somewhat playfully called the "Eightfold Way." In the teachings of the Buddha a "noble eightfold path" was identified, which, if followed, would lead to a cessation of suffering. The parts of the path included "right speech," "right intention," and "right concentration." In Gell-Mann's scheme the cessation of suffering involved a way to classify the bewildering shower of elementary particles, which no one had predicted and which no one understood. In his scheme the number eight played an essential role. Despite their apparent differences, the particles seemed to be classifiable into multiplets—subsets of particles that had predictable characteristics. One of the first multiplets to be identified was an octet of mesons that included the pi-mesons along with new mesons that had recently been discovered. This is what Gell-Mann lectured about.

I was at the lecture and the Chinese lunch that followed it. Because of the presence in the department of T. D. Lee of parity fame, who was a gourmet and knew the chefs of the various Chinese restaurants near Columbia, we could be guaranteed a veritable feast. But during the lecture Robert Serber, who was a very profound but generally quite reserved senior physicist in the department, asked a question. Serber had studied Gell-Mann's paper on the Eightfold Way and had noticed that the simplest representation arising from the theory was a multiplet that consisted of just three particles. He wanted to know if Gell-Mann had considered that. Gell-Mann said that he had but that it led to a conundrum. He said something about it and then we adjourned for lunch. During lunch there was a furious and very fast-paced discussion, which I watched as if it were some sort of intellectual ping-pong game. I shall describe what it was about.

Let us suppose that we have such a triplet. The objective would be to use the three particles as building blocks out of which to construct all the elementary particles. Such an activity had already been tried. Fermi and Yang, for example, imagined that the pi-mesons were composites of the neutron and proton and their antiparticles—as with the negatively charged pi-meson, which was to be a composite of a neutron and an antiproton. Once the strange particles began appearing, however, it was clear that this model had to be generalized. The Japanese physicist Shoichi Sakata introduced a model in which all the particles were to be made up of three particles that were thought somehow to be more elementary: the neutron, the proton, and the newly discovered Λ^0. While it was possible to make all the particles then known out of these three, the scheme was not really satisfactory. In the first place it was not clear why these three chosen particles were any more elementary than any of the others. Furthermore, the scheme predicted new particles that were never found. Moreover, it was not compatible with the Eightfold Way, which seemed to work. Hence Gell-Mann had to start from scratch.

There was no reason why the Eightfold Way triplet had to include any of the known particles. Indeed, as the Sakata model showed, it was better if it didn't. So Gell-Mann introduced three hypothetical particles, which he called up (u), down (d), and strange (s). The up and down objects had a strangeness of 0, while the strange objects had a strangeness of −1. Now we can begin constructing the known particles. The proton, for example, is uud, while the neutron is udd. The Λ^0 is uds. You can see at once a problem, or at least a puzzle: What electric charges should these particles have? All elementary particles ever observed had charges that were integer multiples of the charge of the electron. For the proton the integer is 1, while for the neutron it is 0. If you try to do this for the various composites made out of these hypothetical building blocks, however, it does not work. Suppose, for example, you give the u unit a positive charge. Then to get the proton charge from uud you must give the d unit a negative charge. But that gives the neutron udd the wrong charge. Integer charges simply didn't work. What to do?

Gell-Mann was driven to try something that was either crazy or very audacious: he assigned these particles fractional charges! The u was assigned a charge of $\frac{2}{3}$, the d $-\frac{1}{3}$, and the s $-\frac{1}{3}$. Now the scheme worked, but at the cost of introducing a type of particle that had never been observed. This was what we discussed that day at lunch. Where were these particles? If they existed, they would stand out like a sore thumb. Soon after, a worldwide search began. The most interesting experiments were carried out by the late Peter Franken. He noted that most of the Earth's surface is ocean, so the most likely place to find these objects was in seawater. Oysters concentrate seawater, so the way to look for these objects, he reasoned, was to study oysters. Even if he didn't find them, he could always eat the oysters. Neither he nor anyone else ever found any trace of these fractionally charged particles, however.

In the meantime, Gell-Mann had come up with a name. He has said that when he was first working on these objects he did not have a name but rather a sound. He is an avid bird-watcher, so the sound he came up with was something like "quack." He may have had in mind that in proposing these quack particles he might be regarded as one himself. He also had a considerable interest in linguistics, so the kind of wordplay you find in *Finnegans Wake* was very appealing to him. While trolling through the book he hit on the sentence "Three quarks for Muster Mark," and quacks became quarks. Thus a new term was introduced into the language—or nearly new, for it turns out that quark is also the name for a common German cheese made out of sour milk (see Figure 14).

Unknown to Gell-Mann, or indeed any of the rest of us at the lunch, a graduate student at Caltech, George Zweig, had come up with an identical scheme. Zweig called his fractionally charged particle aces and showed how all the known particles could be built up from them. Ironically, Zweig had been a student of Gell-Mann's, but Gell-Mann had taken a sabbatical, so he had turned Zweig over to Richard Feynman. Feynman did not take many students, and Gell-Mann had had to reassure him that Zweig was competent. Feynman was not that interested in the model, and so Zweig wrote his thesis on other elementary particle topics. But then Zweig went to CERN in Geneva, where he took the opportunity to write up his model. He wanted to publish it in an American journal such as the *Physical Review*, but there was a problem: at the time, these journals had high page charges, and the author or the author's institution had to pay them in order to publish an article. On the other hand, the European journals frequently did not have page charges, and a few even paid the author a small amount. The people at CERN refused to pay the large page charges in order for Zweig to publish in an American journal, and Zweig refused to publish in a European one, so his work came out as two long internal CERN reports. It took some time before what he had done was recognized. A bit later Feynman nominated both Gell-Mann and Zweig for Nobel Prizes (at that point Gell-Mann already had one), though nothing came of this proposal.

It fairly rapidly became clear to physicists that free quarks were not going to be found. Hence, making a virtue out of necessity, they invented a dynamics that would permanently confine the quarks within the particles they were the constituents of. In this dynamics, quarks exchange particles called gluons. This gluon dynamics has a very special characteristic. I have pointed out before that the forces that we are familiar with—the electric, the strong, and even the gravitational—fall off with the distance that separates the objects. In the case of the strong pion force, this fall-off is very rapid indeed. But the gluon dynamics produces a force that has just the opposite property. It is more like stretching a rubber band, where the restoring force becomes greater the more you stretch. Of course, you can stretch a rubber band enough that it breaks, but not so with quarks. The confining force simply gets stronger. There is no escape. The quarks are imprisoned forever.

This raised the question—at least for some people—of the sense in which one can say that quarks "exist." When I heard this discussion I was immediately reminded of similar discussions that had occurred at the end of the nineteenth century and the beginning of the twentieth as to whether atoms existed; atoms too had not been observed. As I mentioned in the Introduction, there were two kinds of atoms—the physicists' atom and the chemists' atom. The chemists' atom is what's called a visualizing symbol—you can make a little diagram for H_2O without knowing anything about the mass or size of the hydrogen and oxygen atoms. The diagram is a kind of bookkeeping device. But physicists want to know precisely the masses and sizes of the atoms. A chemist might well argue that atoms in their sense existed, while physicists might have expressed reservations. With the quarks it was even worse. The free quark was in principle unobservable. Maybe it was a pure visualizing symbol. Gell-Mann sort of punted. He said that the quarks might simply be a theoretical crutch—"current quarks," he called them—that guided physicists in formulating theories, after which the quark underpinning could simply be discarded. But they might also be "constituent quarks," out of which particles are actually constructed. This matter was resolved when it turned out that the confined quarks could actually be observed.

To give you an idea of how such a thing was possible, I want to recount a little fable that George

Gamow presented in one of his popular books as to how Rutherford and his collaborators discovered the nucleus buried deep in the interior of the atom. Ganow imagined some country that had rebels it was trying to deal with. These rebels needed cannonballs. The method they used to try to smuggle them in was to hide them in bales of cotton. Customs officials could open up these bales by brute force, causing all kinds of damage. But they hit on a cleverer method: they would shoot pistols into the bales. If the bullets went right through, the customs officials could be sure that nothing solid was inside, but if the bullets bounced back or were widely deflected, they had better look for the cannonballs. In Rutherford's experiment the "bales" were thin foils of gold and the "bullets" were alpha particles. Although Rutherford was quite confident that nothing unexpected was going to happen and that the alpha particles would sail though the foil undeflected or at least very little deflected, some bounced back. The alpha particle had struck something hard in the atom—its nucleus. That was the idea for how to find the confined quarks hidden in the particles.

In the 1960s a new accelerator was built at Stanford University—the Stanford Linear Collider.

Unlike a cyclotron, in which particles are circulated in circles, in this machine electrons are accelerated to an enormous energy down a two-mile straight path. These electrons are the rifle bullets in Gamow's image, and the protons they collided with were the cotton bales. Physicists skeptical about quarks would have guessed that the electrons would go more or less straight through the proton. But this is not what happened. The results of the experiments showed that the electrons had struck objects inside the proton. Moreover, these objects had the fractional quark charges. Thus it was hard to argue that quarks did not exist. All the mesons known at the time could be built readily out of the three quarks, u, d, and s. But when it came to the baryons, there was a problem.

Once again Wolfgang Pauli comes into our story. After Bohr introduced the notion of the quantized electron orbits around the nucleus, he wanted to use this picture to account for the atoms we actually observe. He proposed in 1922 that the electrons were arranged around the nucleus in layered "shells." The first shell could contain at most two electrons, which corresponded to atomic helium. The next filled shell could contain at most eight, and next eight, the next eighteen, and so on. This way the periodic table got built up. The scheme seemed to work, but Bohr could offer no explanation for these numbers. Pauli could, however.

Pauli's work depended at its base on the idea that all electrons are the same. It is true that when you observe an electron it might be in a different place or might have a different momentum or angular momentum, but it would always be an electron. It would always have the same mass, the same spin, and the same charge. It carries no special mark of identity. Now suppose you have two electrons in a state with no orbital angular momentum, like the ground state of the helium atom. But they do have spins. The spins can be pointing in the same direction or in opposite directions. If the spins are pointing in the same direction, the two electrons can be exchanged and you will never know the difference. This expresses itself in the quantum theory in the behavior of the function that describes this situation. If you exchange the variables for the two identical electrons, the function can go into either itself or its negative. Both are acceptable, since the physics depends only on the square of the function. There are particles called Bose-Einstein particles—the pion is an example—which have integer spins. For these particles the exchange produces no change in the function. Then there are particles called Fermi-Dirac particles—the electron is an example—where the exchange produces a minus sign.

Next, suppose we have two electrons in exactly the same state. If we exchange them, we get the minus sign. This says that the function is equal to its negative, which is possible only if the function is 0. This is the Pauli exclusion principle. No two electrons can be in exactly the same state. We can apply this to the ground state electrons of helium, which have only spin angular momentum, as their orbital angular momentum is 0. There are only two possibilities: one electron has spin up and the other has spin down, or vice versa. If you try to add a third electron, its spin will match one of the others, which Pauli forbade. Hence the first electron shell closes at two, and the third electron in the ground state of lithium starts a new level.

We shall now see how this applies to the quarks. The quarks have spin ½, which is necessary because when you combine three to make a proton (uud) the resultant particle must have spin ½. There is no violation of the Pauli principle here, since we can put the u quarks in different spin states. Table 2 is a table of baryons showing their quark content. If you look back at the meson table (Table 1), you will see the quark contents. (You will also notice that there are new quarks we have not yet spoken about, ones labeled c, t, and b. We will come back to these.)

Most of them have quark contents that are perfectly consistent with the Pauli exclusion principle. But a few are black swans. I call your attention to the Ω^-, which was found in Brookhaven in 1964. It has spin $\frac{3}{2}$ and strangeness -3. It is made up of three s quarks, which Pauli's principle forbids. I will return to this, but first I want to finish some business with the Eightfold Way.

Gell-Mann spent the fall and spring semesters of 1959–1960 in Paris. So did I, having won a National Science Foundation fellowship. The previous spring he had come to Princeton and had paid a visit to the Institute for Advanced Study, where I was a visitor. I had been working on another of his ideas, so we talked. He asked me what I was doing the next year, and I said that I was going to Paris. He suggested that we might work together. "Stick with me, kid, and I will put you on Broadway," is what he said. As it happens, we are almost the same age and graduated from the same high school, although he was three years ahead of me and went to Yale at the age of fifteen. We saw a lot of each other in Paris, and I was able to follow him to some extent as he struggled to understand the particles.

I use "to some extent" advisedly. All during that time Gell-Mann was studying some mathematical

Table 2. Baryons.

Particle	Symbol	Makeup	Rest Mass (MeV/c^2)
Proton	p	uud	938.3
Neutron	n	ddu	939.6
Lambda	Λ^0	uds	1115.6
Sigma	Σ^+	uus	1189.4
Sigma	Σ^0	uds	1192.5
Sigma	Σ^-	dds	1197.3
Delta	Δ^{++}	uuu	1232
Delta	Δ^+	uud	1232
Delta	Δ^0	udd	1232
Delta	Δ^-	ddd	1232
Xi Cascade	Ξ^0	uss	1315
Xi Cascade	Ξ^-	dss	1321
Omega	Ω^-	sss	1672

objects that arose in the theory. He kept saying that they commuted like angular momenta. I understood that much. Ordinary numbers A and B commute, in that $AB=BA$. In general, however, the quantities that represent observables in quantum mechanics do not commute. For example, the quantities that represent the components of the

Spin	Strangeness	Lifetime (seconds)	Decay Modes
$\frac{1}{2}$	0	Stable	
$\frac{1}{2}$	0	881.5	$p e^- \bar{\nu}_e$
$\frac{1}{2}$	−1	2.6×10^{-10}	$p\pi^-$, $n\pi^0$
$\frac{1}{2}$	−1	0.8×10^{-10}	$p\pi^0$, $n\pi^+$
$\frac{1}{2}$	−1	6×10^{-20}	$\Lambda^0\gamma$
$\frac{1}{2}$	−1	1.5×10^{-10}	$n\pi^-$
$\frac{3}{2}$	0	0.6×10^{-23}	$p\pi^+$
$\frac{3}{2}$	0	0.6×10^{-23}	$p\pi^0$
$\frac{3}{2}$	0	0.6×10^{-23}	$n\pi^0$
$\frac{3}{2}$	0	0.6×10^{-23}	$n\pi^-$
$\frac{1}{2}$	−2	2.9×10^{-10}	$\Lambda^0\pi^0$
$\frac{1}{2}$	−2	1.693×10^{-10}	$\Lambda^0\pi^-$
$\frac{3}{2}$	−3	0.82×10^{-10}	$\Xi^0\pi^-$, $\Lambda^0 K^-$,

. . .

orbital angular momentum do not commute with each other. But they obey a very simple commutation relation. If you commute the components in the x and y directions, say, the result is proportional to the component in the z direction. The constants of proportionality are very well defined. But what this had to do with the elementary particles was beyond me.

It shouldn't have been, because I had been a math major before I switched into physics, and among the things that I had studied were what are called Lie algebras (after Sophus Lie, who invented them). What Gell-Mann was saying, although he did not realize it, was that his quantities formed a Lie algebra. There was no application of this work in classical physics. Nonetheless, it happened that the French mathematician Elie Cartan in his thesis in the 1920s had made a classification of these algebras. Gell-Mann was a visiting professor at the Collège de France, and any of the mathematicians he regularly had lunch with could have told him that he was trying to find a Lie algebra that had representations bearing some resemblance to the observed elementary particles. In the fall Gell-Mann returned to Caltech and conferred with an assistant professor of

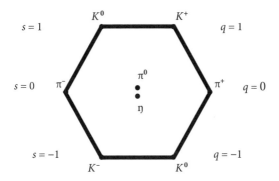

Figure 15. The meson octet.

mathematics there named Richard Block. Block explained to him what he was trying to do, and very shortly Gell-Mann found the Lie algebra that worked. Its technical name is SU(3). It has representations in triplets, octets, and decuplets, as well as higher-dimensional multiplets. The octet was immediately identifiable. It consisted of the three pions, the four K-mesons, and a new meson called the eta, which had been discovered at the Bevatron in Berkeley in 1961. Figure 15 shows the octet diagram. You must admit that it is a thing of beauty.

Now we can return to the Ω^-. First, how is it classified, and then how do we deal with the Pauli problem?

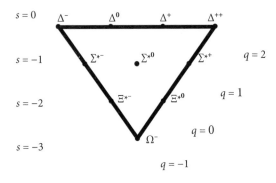

Figure 16. The baryon decuplet.

Figure 16 shows the baryon decuplet. Some of the particles are already familiar to you, and with the exception of the Ω^-, all the others were known at the time Gell-Mann made his proposal. He was able to tell the experimenters what to look for, and as I mentioned, the Ω^- was found at Brookhaven in 1964. In 1969 Gell-Mann was awarded a well-deserved Nobel Prize.

This still leaves us with the Pauli problem, however. Theoretical physicists earn their keep by being ingenious, and what was needed was to invent some distinguishing characteristics for the quarks. Thus the notion of color was introduced. It was actually invented before it was needed. A theorist named

O. W. Greenberg had as early as 1962 introduced a kind of statistics that he called "parastatistics." It was an intermediate step between the Bose and Fermi statistics and was an interesting idea, but at the time it had no applications. Once the dilemma with the quarks was discovered, however, the theorist Yoichiro Nambu (of whom we will hear more later) and his graduate student Moo-Young Han saw how to apply Greenberg's idea to the quarks. You would take a flavor, such as up, and render it in triplicate by associating it with three different colors, one for each of the three quarks. (Color is simply a mnemonic for the characteristics, of course. Which three colors were actually chosen depended on the nation of origin of the physicists. Americans usually chose red, white, and blue.) When you write the three s quarks for the Ω^- you choose one of each color and hence avoid a conflict with Pauli. This triplication shows up in experiments of the Stanford type, so it seems real.

As the years went on new particles were discovered and new quarks were needed. There are now six flavors. Table 3 shows the quarks that are needed to make up the known particles. The new quarks are bottom (b), charm (c), and top (t).

The baryon number of a neutron or proton is 1, which is why each quark has a baryon number of $\frac{1}{3}$. The masses are given in terms of electron volts (see Appendix 1 for an explanation). The masses of the up and down quarks are not precisely known.

✳

In 1970, when I was a visitor in the Theory Division at CERN, there was an announcement that Sheldon Glashow was coming to give a talk. I had first met Glashow in the mid-1950s, when he came to Harvard from Cornell to do his graduate work. At the time there was a windowless and usually smoke-filled room in the basement of the physics building

Table 3. The quarks needed to make up the known particles.

Name	Symbol	Mass (Mev/c^2)	Spin	Baryon Number
Up	u	1.7 to 3.3	$\frac{1}{2}$	$+\frac{1}{3}$
Down	d	4.1 to 5.8	$\frac{1}{2}$	$+\frac{1}{3}$
Charm	c	1,270	$\frac{1}{2}$	$+\frac{1}{3}$
Strange	s	101	$\frac{1}{2}$	$+\frac{1}{3}$
Top	t	172,000	$\frac{1}{2}$	$+\frac{1}{3}$
Bottom	b	4,190	$\frac{1}{2}$	$+\frac{1}{3}$

that was occupied by the people who were trying to get their Ph.D.'s in theoretical physics, and thus the room had an atmosphere of angst. I had my degree, so I avoided going down there unless I wanted to see a friend. When I did visit, I noticed Glashow, a tall fellow who seemed to be having a very good time. Indeed, every time I saw him he seemed to be playing a game. The African strategy game wari was a favorite. If I'd had to make a guess, I would have thought that Glashow would be the last person in this group to win a Nobel Prize. (It took him until 1979, when he received the Nobel for work that he had started for his Ph.D. thesis with Schwinger.) At CERN years later I decided to go to

Electric Charge	Strangeness	Antiparticle	Antiparticle Symbol
$+\frac{2}{3}$	0	Antiup	u^c
$-\frac{1}{3}$	0	Antidown	d^c
$+\frac{2}{3}$	0	Anticharm	c^c
$-\frac{1}{3}$	-1	Antistrange	s^c
$+\frac{2}{3}$	0	Antitop	t^c
$-\frac{1}{3}$	0	Antibottom	b^c

his lecture, thinking that whatever he had to say was bound to be entertaining.

What he was concerned with was an anomaly in the decays of the mesons. If you look at Table 1, in Chapter 6, you will find some mesons labeled D in the far left-hand column. If you look at the lifetime for the decay of these mesons, which is given in the far right-hand column, you will see that it is longer than the others by a factor of some ten billion. Glashow and his collaborators John Iliopoulos and Luciano Maiani had come up with an explanation: there must be a new conservation law that was at least approximately observed. Recall we had a similar anomaly when we considered the decay $\mu \rightarrow \epsilon + \gamma$, which does not occur at the rate one thinks it should. In this case the conservation law was the lepton number. The new conservation law was named the conservation of charm. (Glashow invented the term *charm*, although why he chose this word I do not know.) The D-meson contained one of these c quarks and hence was "charming." But it decayed into charmless particles, hence the longer lifetime. I could not see anything wrong with this, except that it struck me as like trying to kill a gnat with a sledgehammer. So I decided to

forget it. This state of indifference persisted until
1974. On November 11 of that year two widely sepa-
rated laboratories announced the same extremely
interesting discovery of a new meson that had a
somewhat longer lifetime than the others.

The West Coast experiments were done at the
Stanford Linear Accelerator by a group headed by
Burton Richter, and the East Coast experiments were
done at Brookhaven by a group headed by Samuel
Ting of MIT. The question was what to call this new
meson. In California, Richter wanted to call it the
SP, after Stanford's SPEAR accelerator, though he
was persuaded to call it the psi, since it had the letters *sp* reversed; in New York, the preferred name
was J, since the Chinese characters for Ting's name
are 丁肇中 and the first one looks sort of like a *J*. So
the two names were combined and the particle was
called the J/Ψ. Its properties are given in Table 1.
One might think that this meson could be made up
out of quark-antiquark pairs of the usual type. But
it decays into K-mesons, which are also made up of
quark-antiquark pairs of the usual type. Yet this decay is suppressed, so the question is why. The natural
answer is to suppose that the J/Ψ is made up of a
charm-anticharm pair. Now the decay into particles

that have no charmed quarks is suppressed and the result is explained to everyone's satisfaction. The charm-anticharm pair can have different energy levels. The J/Ψ represents the lowest of these levels. The higher levels correspond to more massive objects, some of which have been observed. This system of charm-anticharm quarks is called charmonium.

There are two more quarks that we know about—the top and the bottom. The bottom quark was discovered at the Fermi Lab near Chicago by a group led by Leon Lederman. Using protons from the accelerator, they created a new meson, the upsilon (see Table 1). It has a somewhat longer lifetime than expected, so, using an argument similar to the one we used about the charmed quark, the existence of a new quark—the bottom quark—was accepted. The top quark is so massive that it is barely observable even in the largest accelerators, but the evidence appears convincing. Are there more lurking at higher energies? Perhaps.

*

I want to close this chapter with a little philosophical homily—a comparison of physics to biology. It involves the aforementioned George Gamow. Gamow, who was a larger-than-life Russian-born eccentric, was also one of the most imaginative physicists of the twentieth century. For example, he had worked out some of the physics of the Big Bang long before it was called that, and long before anyone else was much interested. He also persuaded Rutherford to establish the first particle accelerator at Cambridge University. In the spring of 1953 he went to Berkeley to give a lecture. While he was there the physicist Luis Alvarez showed him two articles in *Nature* written by James Watson and Francis Crick. They were the articles in which the double helix model of DNA was presented.

When Gamow read them, he realized what was missing. DNA is a long strand—in fact, two long strands—made up of bases, which I will simply refer to as T, A, G, and C. These bases are strung together like beads in a linear chain. What Gamow understood, as did Watson and Crick, was that all genetic information must be in these bases. Gamow had no idea how the information was gotten out; what he was concerned with was how the information was coded. He knew that this information specified the amino acids that went into the manufacture of proteins. He realized that there were a limited number of amino acids that were used (there are twenty, in fact), and he reasoned that some arrangement of the letters of the bases must be the code for a particular amino acid. He realized that two letters would be too few and four needlessly many, and hence concluded that there would be three-letter combinations, such as TGG or AGC, which meant there were $4 \times 4 \times 4 = 64$ possibilities. He decided that the genetic code must be "elegant" the way the laws of physics are, with no duplication—there could be only one "word" per amino acid. He produced a very ingenious model that—apart from the three-letter combinations—was entirely wrong, for he had forgotten evolution. In evolution, elegance is not what counts—only adaptability does. In fact, the genetic code could be described as "degenerate," in that more than one word can specify an amino acid and some words are signals for starting or stopping a sequence. Unlike physics, the genetic code cannot be deduced. Someone who knew all the laws of nature at the beginning of time could predict all the atoms that could ever exist, and someone who knew all the quarks could predict all the baryons and mesons that could ever exist. But that knowledge could never predict the existence of a giraffe.

Part III
Pastels

Figure 17. The ring of the Large Hadron Collider at CERN. It is well underground. The ring has been superposed by the artist to show where the tunnel goes. Photo courtesy of the European Organization for Nuclear Research (CERN).

8
The Higgs Boson

It was understood from the beginning that the Eightfold Way was only an approximate symmetry. This is evident from the masses of the particles. The pi-mesons are placed in an octet that also includes the strange K-mesons. The mass of the K-mesons is more than three times the mass of the pi-mesons. If the symmetry were exact, all the particles in the octet would have the same mass. But when you look at the properties of these particles it is not clear at all that they have anything to do with each other. That is why finding the symmetry that unites them was so difficult.

The kind of symmetry breaking that is manifested in the Eightfold Way is as old as the quantum theory itself. It was brought into the theory by the mathematician Hermann Weyl and the physicist Eugene Wigner. Wigner applied it to atomic spectra. The electrons outside the nucleus of atoms are located on orbits. The quantization of these orbits was, as we have seen, the great work of Niels Bohr. If the atom is not disturbed, the electrons find their way to the orbits of the least energy—the ground states. If the atom is excited by heat or an electric shock, the electrons jump to the higher orbits. But once the disturbance is over they go back to the ground state, emitting radiation (which is related to the energy loss as the electrons jump back down). This produces beautiful and characteristic atomic spectra. The atomic spectrum of an element is like a fingerprint that identifies that element. That is how helium was first discovered in the atmosphere of the Sun. These lines reflect the symmetry of the theory that describes these energy levels. However, this symmetry can be compromised by putting the atoms in a magnetic field. This breaks the symmetry in a very well-defined way, and as a consequence, what was a single spectral line becomes a multiplet, as we have already discussed.

The number and intensity of these lines can be predicted, as they reflect the symmetry that was broken. This kind of Wigner-Weyl symmetry breaking is at play in the Eightfold Way. One adds a small term that breaks the symmetry, so the multiplet masses get split. If one does this with tact, then the resulting masses are related. Here, for example, is Gell-Mann and Okubo's mass formula for the familiar baryons:

$$2(m_N + m_\Xi) = 3m_\Lambda + m_\Sigma$$

The reader is invited to test this by putting in the empirical masses and taking the average mass for the proton and neutron and the average mass for the Σ. You will find that the agreement is quite good, which offers additional confidence about the scheme.

When I was learning quantum mechanics in the late 1950s we were taught Wigner-Weyl symmetry breaking. But in 1961 Yoichiro Nambu and his student Giovanni Jona-Lasinio published two papers that changed everything. This was an entirely different kind of symmetry breaking.

In his 2008 Nobel Lecture Nambu gave an example of this type of symmetry breaking. Suppose you have a straight rod that is elastic and you stand it vertically. It looks the same from any horizontal direction. The symmetry is perfect. But if you squeeze it down, it will bend in some direction and the symmetry is gone. An example that I like is to take the same rod and balance it on a point. Quantum mechanics tells us that this balance can never be perfect. There is always some quantum uncertainty in the angle it makes with the ground. Hence it will tip over, but it is equally likely to fall in any direction. Once it does this, the symmetry is gone. Putting the matter somewhat more abstractly, equations can show symmetries, but you are not forced to choose solutions that respect these symmetries. This kind of symmetry breaking is characterized as spontaneous, to distinguish it from the Wigner-Weyl type, in which a symmetry-breaking term is added to the equations. In the spontaneous type nothing is added to the equations.

Nambu applied his idea to magnets, crystals, and superconductors as well as to the theory of elementary particles. When he studied a model of the last, an uninvited guest appeared. But before going further, let me reintroduce a term of art—*boson*. In the mid-1920s an Indian physicist named Satyendra Nath Bose (pronounced "bosh") introduced a kind of statistics that applied to particles such as the pi-meson and K-meson. He applied it to the photon as well, and sent his paper to Einstein, who made some corrections and had it published, translating it himself from English into German. Particles of this kind are called bosons. The other class of particles, which includes the electron, the proton, and the neutrino, are called fermions, after Fermi, who did pioneering work on their statistics. These are the ones, as we have seen, that obey the exclusion principle.

The uninvited guest in Nambu's model was a boson of zero mass. At the time, it was known that the photon had zero mass, and it was thought that the neutrinos had zero mass (which they don't), but no boson of zero mass apart from the photon was known, nor is one known to this day. At first it was thought that this might be an artifact of Nambu's model, but in 1961 the British-born physicist Jeffrey Goldstone found another model with these particles, which are now called Nambu-Goldstone bosons. The feeling grew that these particles were a blight that came with this spontaneous symmetry breaking and hence might not have much relevance to the world of elementary particle physics.

The breakthrough came in 1964 from physicist Walter Gilbert. He had been appointed an assistant professor of physics in 1959, but in 1964, the year in which he wrote his groundbreaking paper, he became an associate professor of biophysics at Harvard. He won the 1980 Nobel Prize for chemistry and went on to round out his remarkable career by cofounding Biogen. There is nothing wrong with the arguments of Goldstone and Nambu, but they are not general enough. Gilbert analyzed the assumptions and noted that they did not apply to electrodynamics. This was a wonderfully liberating discovery, and it was first fully exploited for elementary particles, at least in print, by the English physicist Peter Higgs. In this business there is much at stake—potential Nobel Prizes, for example—in the assignment of credit. I was not there, but I have Peter Higgs's papers in front of me, and I think I know what they say and what they do not say. Higgs considered a theory, a form of electrodynamics, that began with a massless photon and some bosons that coupled to each other in a special way and to the photon. This theory enjoys an apparent symmetry that is broken spontaneously. If it were not broken, the photon would remain massless. But with the breaking, something miraculous then happens: the massless "photon" acquires a mass, and the theory acquires a massive boson, which is now universally called the Higgs boson. You have to see the mathematics to believe it.

The first person to take advantage of all of this in a substantial way was Steven Weinberg. To explain what he did, I have to back up some. Looked at on its face, as we have seen, there are four basic forces. I put it this way because the holy grail is to produce a theory of everything in which these forces will be unified. But we should understand what forces we are talking about. The strong force is what holds the quarks in their particle confinements. It is mediated by the exchange of gluons—massless photonlike objects that carry the color charge but not an electric charge. These gluons change the color of the quarks. In the theory the gluon force has the property I mentioned before of increasing with distance, so the quarks are confined. Next in strength is the electromagnetic force, which is mediated by the exchange of photons. I will skip over the weak force for the moment and mention the weakest of them all, which is the gravitational. It is mediated by the exchange of gravitons. They have never been observed as free particles, and some have conjectured that they never can be. (I will come back to this in the next chapter.) It may seem odd to think of gravitation as being the weakest force, since it is the force we deal with on a daily basis. But it is easy to compare the attractive force of gravity between two protons and the repulsive electric force. Gravity is some 10^{36} times smaller, which is why we can ignore it in our considerations about nuclei.

The weak force is responsible for reactions such as beta decay. As I mentioned in Chapter 2, it was Fermi who developed the first real theory. He tried to model it on electrodynamics, but there is a big difference. In Fermi's theory the interaction was in effect mediated—although he did not put it this way—by a particle whose mass was infinite. The force had zero range, meaning the particles involved had to be on top of each other when they interacted. The theory explained a lot but came to seem out of joint with the other forces, which were mediated by particles with reasonable masses, such as the pion. Hence the physicists invented particles to mediate this interaction that were supposed to be very heavy but had finite masses. They were three in number—two charged ones, called W mesons (the *W* was for *weak*), and a neutral one, called the Z meson (for reasons unclear to me). Various properties were conjectured. Remarkably, both the W and the Z were observed in 1983 at CERN, using what was called the Super Proton Synchrotron (SPS). The W mesons were found to have masses that were something like ninety times the proton mass; the Z is slightly more massive.

The Z paper's list of authors and their institutions is instructive:

G. Arnison, A. Astbury, Bernard Aubert, C. Bacci,
G. Bauer, A. Bezaguet, R. Bock, T. J. V. Bowcock,
M. Calvetti, P. Catz, P. Cennini, S. Centro,
F. Ceradini, S. Cittolin, D. Cline, C. Cochet,
J. Colas, M. Corden, D. Dallman, D. Dau,

M. DeBeer, M. Della Negra, M. Demoulin,
D. Denegri, A. Di Ciaccio, D. DiBitonto,
L. Dobrzynski, J. D. Dowell, K. Eggert,
E. Eisenhandler, N. Ellis, P. Erhard, H. Faissner,
M. Fincke, G. Fontaine, R. Frey, R. Fruhwirth,
J. Garvey, S. Geer, C. Ghesquiere, Philippe Ghez,
W. R. Gibson, Y. Giraud-Heraud, A. Givernaud,
A. Gonidec, G. Grayer, T. Hansl-Kozanecka,
W. J. Haynes, L. O. Hertzberger, C. Hodges,
D. Hoffmann, H. Hoffmann, D. J. Holthuizen,
R. J. Homer, A. Honma, W. Jank, G. Jorat,
P. I. P. Kalmus, V. Karimaki, R. Keeler, I. Kenyon,
A. Kernan, R. Kinnunen, W. Kozanecki, D. Kryn,
F. Lacava, J. P. Laugier, J. P. Lees, H. Lehmann,
R. Leuchs, A. Leveque, D. Linglin, E. Locci,
J. J. Malosse, Thomas W. Markiewicz, G. Maurin,
T. McMahon, J. P. Mendiburu, M. N. Minard, M.
Mohammadi, M. Moricca, K. Morgan, H. Muirhead,
F. Muller, A. K. Nandi, L. Naumann, A. Norton,
A. Orkin-Lecourtois, L. Paoluzi, F. Pauss, Giovanni
Piano Mortari, E. Pietarinen, M. Pimia, A. Placci,
J. P. Porte, E. Radermacher, J. Ransdell, H. Reithler,
J. P. Revol, J. Rich, M. Rijssenbeek, C. Roberts,
J. Rohlf, P. Rossi, C. Rubbia, B. Sadoulet, G. Sajot,
G. Salvi, G. Salvini, J. Sass, J. Saudraix, A. Savoy-
Navarro, D. Schinzel, W. Scott, T. P. Shah, M. Spiro,
J. Strauss, Jonathan Mark Streets, K. Sumorok,

F. Szoncso, D. Smith, C. Tao, Graham Thompson, J. Timmer, E. Tscheslog, J. Tuominiemi, B. Van Eijk, J. P. Vialle, J. Vrana, V. Vuillemin, Horst D. Wahl, P. Watkins, J. Wilson, C. Wulz, G. Y. Xie, M. Yvert, E. Zurfluh

Aachen, Tech. Hochsch. & Annecy, LAPP & Birmingham U. & CERN & Helsinki U. & Queen Mary, U. of London & Collège de France & UC, Riverside & Rome U. & Rutherford & Saclay & Vienna U. & NIKHEF, Amsterdam & Wisconsin U., Madison & Kiel

Somewhere in the middle of this list of 137 names the reader will find that of Carlo Rubbia, who was the conductor of this giant orchestra. He designed the experiment and got the Nobel Prize. I am not sure how the 136 other people felt about this, but they must have known what their roles were. It is, of course, absurd to compare this situation with that of Rutherford and his two colleagues discovering the atomic nucleus more or less on a tabletop. Perhaps a closer comparison is with the Harvard Cyclotron. That machine was run by the experimenters—in the main, students and their professors. There might have been three or four people involved. I was occasionally asked to pile lead bricks for shielding or even to sew targets. On the CERN accelerators, by contrast, the experimenters get nowhere near the machine, which is run by professional engineers. The detectors are the size of buildings, and the results can be analyzed only by computers and more technologists. The cost runs to billions and is shared by the countries that contribute their taxpayers' money to CERN. This is what it takes to do this kind of particle physics. I am constantly amazed that the money is forthcoming.

Physicists had already assumed these particles existed, and had been constructing theories with them for many years. However, these theories had an issue: they did not make sense. Put more concretely, they produced infinities that could not be gotten rid of. To explain this, I have to return to the electron and photon.

A quantum world in which there are only electrons and photons is described by the theory of quantum electrodynamics. The theory had its origins at about the same time as quantum mechanics was invented, in the late 1920s. The pioneers were a familiar cast of characters that included, among others, Dirac, Heisenberg, and of course Pauli. This theory could describe the emission and absorption

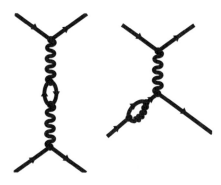

Figure 18. Corrections to the diagram shown in Figure 8.

of photons and things such as the annihilation of an electron-positron pair into two gamma rays. You could readily reproduce the formula Rutherford had derived for the scattering of a pair of charged particles. The quantum electrodynamic diagram that leads to this is one that we have seen before (see Figure 8 in Chapter 4).

But what quantum electrodynamics claimed to be able to do was to calculate the corrections to this diagram. Two examples are shown in Figure 18.

These diagrams should produce small corrections, but instead they turned out to produce infinities.

All during the 1930s theorists agonized over this dilemma. One problem was that there was no experimental guidance. The experiments were not accurate enough to detect departures from the non-corrected expressions. This changed drastically after World War II, however. During the war a number of physicists had worked on radar and other developments that had taught them new skills, which they later applied to experiments on these effects. Two of special importance were done at Columbia University. One, in 1947, involved the collaboration of Willis Lamb and Robert Retherford and focused on the energy levels of the hydrogen atom. The standard theory at that time, which used the Dirac equation, predicted that two of the lower levels should have the same energy. There had been prewar suggestions that this might not be the case and that the higher-order effects of quantum electrodynamics might be responsible. But the experiments had not been done and the theory produced an infinite correction. In their work, however, Lamb and Retherford measured with great precision what is known as the Lamb shift. It was perfectly finite and demanded an explanation. The second thing that was measured was a magnetic property of the electron. The spinning electron acts like a tiny magnet. Its strength is measured by what is called its magnetic moment. Again the Dirac theory predicted what this should be, and it is generally given in terms of a dimensionless constant called g. The Dirac theory predicted that g should equal 2, but experiments showed that it departed from 2 by a tiny amount. The recent experimental value (with error) is $2.00231930419922 \pm (1.5 \times 10^{-12})$. Note that it is only two thousandths larger than the value from the Dirac equation. This was again assumed to be a quantum electrodynamical effect, and again, computing it gave an infinite answer. Now there was a real challenge.

This is not the place to go into detail of how the challenge was met. In brief, it was largely the work of Julian Schwinger, Feynman, and Sin-Itiro Tomonaga (the last of whom worked with his small group in isolation in Japan during the war), and Freeman Dyson knitted all of this together in a series of monumental papers. What I will discuss here is the notion of renormalization. When, to take an example, Rutherford scattering is done using only the simplest Feynman diagram, you insert into the diagram a value for the electron charge. This charge is often referred to as the bare charge. But when you take into consideration the corrections, the charge is also corrected. The problem is that this corrected charge is infinite. The measured charge is proportional to the bare charge, and this constant of proportionality is infinite. Renormalization consists of ignoring this fact and simply replacing the bare charge by the measured one. Something similar has to be done for the electron mass. The remarkable thing is that when this is done it renders the terms in the Feynman diagrams finite. Dyson showed that to every order this works. That is why you can calculate, for example, g to such accuracy, namely, $g_{theory} = 2.0023193048$. You are invited to compare this to the experimental value.

A theory in which this procedure can be done with only a finite number of such alterations is called renormalizable. Quantum electrodynamics is the poster child for such a theory. It requires only two such redefinitions. You may well have some queasy feeling about this. After all, the infinities are still there; they have just been shoved under the rug. For a while Dyson tried to show that the full theory was finite, but he found a physical argument to show that it probably isn't. Still, some theories are not even renormalizable; as we will discuss in the next chapter, the theory of quantum gravity is one. But so is the theory of weak interactions, in which the W and Z mesons are simply put in by hand. It presented the paradox that so long as you did not worry about corrections it worked wonderfully, but when you took these into account you got nonsense. This produced a good deal of desperation physics. And then came Steven Weinberg.

In Weinberg's electroweak theory you begin with a symmetric situation in which both the photons and the weak mesons have no mass. They interact with bosons along the lines of the Higgs model. The symmetry is spontaneously broken and the same miracle happens: the weak mesons acquire a mass and the photon doesn't. The Higgs boson acquires a mass. It all fits together like a Swiss watch. Weinberg conjectured that the theory would be renormalizable, and this was proved in detail by Martinus Veltman and his student Gerardus 't Hooft. They shared the Nobel Prize for this work. But there is a price—or an opportunity, depending on how you look at it. The theory generates a Higgs boson. Without it the whole thing collapses. This launched a monumental experimental quest.

Never have so many physicists spent so much time and money in such a quest. There are three accelerators that have been or are in the hunt—two of them at CERN and one at Fermilab, near Chicago. The oldest of these is the Large Electron Positron (LEP) Collider at CERN, now decommissioned. It showed that the Higgs boson could not have a mass less than about 120 proton masses. But it found no credible evidence for the particle. The experiments at the Tevatron (also now decommissioned) at Fermilab tightened the limit considerably, hence we knew how light the particle could be. The theorists presented convincing arguments as to how heavy it can be—something a little over 100 proton masses. This meant that it was in the energy range accessible to the Large Hadron Collider (LHC), and indeed, on July 4, 2012, two experimental teams at CERN announced very impressive evidence that the particle exists. It is worth discussing the form that this evidence takes.

Figure 17 shows the location of the Large Hadron Collider at CERN. As its name suggests, the Large Hadron Collider collides hadrons—in this case, protons. These protons are moving at speeds only fractionally lower than the speed of light. Collisions at these speeds involve the constituent parts of the protons—the quarks and gluons. It is these particles that merge and make a Higgs boson. The Higgs boson is very unstable and has several possible decay modes. Three of the most interesting are into a W^+ and a W^- or two Z^0s, or, alternatively, into two gamma rays. The W and Z particles can decay into electrons, muons, and neutrinos. Two Z particles can produce two electron-positron pairs, two muon pairs, or a mixture of electron and muon pairs. These four lepton channels are very striking. It is hard to imagine any other source for them besides a Higgs decay. On the other hand, the gamma ray channel is hard to distinguish from the background. It requires a careful analysis to conclude that what is being observed is not background. The fact that there is a two-gamma decay proves that this particle is a boson—a particle with integer spin. Each gamma has spin 1, so two of them must have an integer angular momentum. Hence so must the newly discovered object if angular momentum is conserved in the decay. The two detectors at the LHC, ATLAS and CMS, are claimed to have detected gammas above the expected background. Indeed, both announced that their data were consistent with a particle with a mass of a bit over 130 proton masses. The simplest assumption is that it has spin 0. It can be shown that the fact that it decays into two gammas rules out spin 1. Spin 0 Higgs can have one of two intrinsic parities, pseudo scalar like the pion or scalar as the standard theory would suggest. The difference shows up in properties of the decays. It would appear as if all the evidence is consistent with a scalar Higgs. There is a sort of gold standard that is used in this work, and that is the number of sigmas, or standard deviations, that can be attached to the data. The higher the number of sigmas, the more reliable the data are said to be. For scientists to declare a discovery such as this, the data should have a five-sigma validity. This means that the data are well over 99 percent certain. The teams at the CMS and the ATLAS each reported a sigma close to five, and if all the data are combined, including those from the Tevatron, this gives a sigma of over seven. This was enough for the CERN experimenters to declare victory. They may well be right, but there was so much pressure to discover this particle that it makes me a bit queasy. Figure 19 depicts a simulated Higgs event as it would have appeared in one of the CERN detectors.

Here I repeat a morality tale that I first noted in Chapter 1. Soon after the neutron was discovered, Enrico Fermi decided to become an experimental physicist and to do experiments with neutrons. His

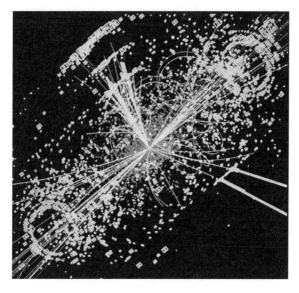

Figure 19. A simulated Higgs event as it would appear in one of the CERN detectors. Photo courtesy of the European Organization for Nuclear Research (CERN).

group in Rome bombarded with neutrons every element they could lay their hands on, working their way up to uranium. Fermi knew what he was going to find with uranium: that he would make a trans-uranic element, one heavier than uranium. This element would be observed not directly but through its decays. Indeed, Fermi found the decays he thought he would, and so he declared victory. In fact, as I have said, what he had discovered—although he did not know it—was nuclear fission. It had nothing whatever to do with transuranic elements. I will keep this story in mind until there are more data from the LHC.

My feelings about the claimed discovery of the Higgs boson are mixed. If they have indeed found it, it will end a chapter in physics. I am reminded of something that a French colleague once said to me. He had had a parameter named after him, and it was much discussed because a theory of the weak interactions hinged on it. Finally the parameter was measured, and it confirmed the theory. I went to congratulate him, and he said that he was unhappy "because now they will speak of it no more." Similarly, I thought that *not* finding the Higgs particle would be more exciting, because new physics would then be needed. I don't believe too many of my colleagues felt this way.

This reminds me of the following story about Einstein. He had just received a telegram with the news that the solar eclipse expeditions had confirmed his general relativity prediction about the Sun bending starlight. He was very pleased with himself and showed the telegram to one of his students, Ilse Rosenthal-Schneider. She asked him what he would have done if the telegram had contained the news that the experiments disagreed with the theory. He replied, "Da könnt' mir halt der lieber Gott Leid tun, die Theorie stimmt doch," which translates as "Then I would have been sorry for the dear Lord. The theory is right."

9
Neutrino Cosmology

The reader may well wonder what the neutrino has to do with cosmology, which is the science of the evolution of the universe from the Big Bang. I have noted that neutrinos can go through light-years of lead hardly interacting. But there is a qualification that I did not state. The neutrinos I had in mind had the sorts of energies found in beta decay. The conditions in the very early universe were entirely different, however.

The Big Bang occurred about 13.7 billion years ago. One can readily get the idea from the name that one is discussing an explosion in which the objects produced expanded into space. But according to present ideas, prior to the Big Bang there was no space-time. What expanded is space-time itself, and the particles came along for the ride. After about 10^{-37} seconds space-time expanded exponentially for a very short time. This exponential expansion, which is called inflation, is necessary to account for some of the regularities we presently observe. This inflationary epoch left its imprint on the blackbody radiation left over from the Big Bang. The curve I showed in Figure 6, back in Chapter 3, is not quite detailed enough to show this, but other experiments do.

When the inflation stopped, we were left with the players in the game that were going to eventually constitute the universe we know today. There are other players we have not yet observed directly such as "dark matter." In the former category are the quarks, antiquarks, and gluons. There are suggestions that there are fewer antiquarks than quarks, which would account for the fact that the universe we know is dominated by particles and not antiparticles. After about a microsecond, things cooled down to where the quarks could be bound by the gluons to make the mesons and baryons. Most were unstable and decayed into something else—the leptons, such as electrons and muons, and the various neutrinos, which were created in the Big Bang along with photons. Particles annihilated with antiparticles, leaving the particle dominance we see today. It was still too hot to allow stable nuclei to be formed; ambient gamma rays would rip them apart.

At about three minutes things had cooled to the point where deuterons could be formed. Two deuterons could fuse to make a helium nucleus, and it is this cosmological helium that interests us. Things were still too hot to allow the formation of atoms; the electrons would have been torn off by the gammas. But after about 370,000 years things cooled off to where atoms began to form. The number of free electrons was greatly diminished. This means that the photons had nothing to collide with and could expand freely with the universe. The cosmic background radiation we observe today is largely a relic left over from this epoch.

If we look around the cosmos, it seems quite irregular. There are stars and galaxies and quasars and black holes. But if you average over everything, the universe is remarkably homogenous. It is not a bad approximation to think of it as matter and energy uniformly spread out. Because of Einstein's equation $E=mc^2$ there is for these purposes no difference between matter and energy. The energetic photons contribute to the matter density of the universe. Because of the uniformity, the expansion is described by a function of the time alone. All the spatial dimensions scale with this function.

Hence to describe the timeline of the expansion we have to know how this function changes with time. The expansion would be free except that gravitation holds it back. Because of gravitation the expansion of the universe decelerated throughout most of its history. However, this expansion is now in an accelerating phase, which means that some antigravitational force is at work. This is usually referred to as "dark energy." Its origin is unclear.

Now to the neutrinos. Under present laboratory circumstances the rate for a reaction such as $e + \nu \rightarrow e + \nu$—the collision of a neutrino with an electron—is extremely small. But it is a very sensitive function of the energy. If the electrons involved are very energetic, as they were in the early universe, the rate is enhanced enormously. The rate goes as the fifth power of the electron temperature. If you raise the temperature by a factor of 10, then the rate increases by a factor of 100,000. At these early universe temperatures this rate was comparable to the rate of collisions of photons and electrons. This means that the neutrinos, electrons, and photons were all at about the same temperature. The early universe was like a blackbody, which has a nearly common temperature for all the objects inside it.

Hence the neutrinos contributed to the rate of expansion of the universe an amount comparable to that contributed by the photons. The more types of neutrinos there are, the greater this contribution. Remarkably, this is something that can be measured. The tool is the cosmic helium.

At the present time, most of the cosmological matter we observe is in the form of protons. The next in total mass is helium. The ratio of the mass of the observed cosmic helium to the total mass, mainly protons, is about 0.23. There is a small contribution to the total mass from cosmological deuterons, but, as I said, most is in protons. The object of theory is to explain this number. To make helium you need neutrons. But there are effects that deplete neutrons. The simplest to explain is the instability of the neutron. The neutron decays in just under fifteen minutes. There is nothing that can be done to change this. But the more rapidly the universe is expanding, the cooler it becomes before this decay time is reached. Hence there will be more neutrons around at the crucial temperature, forming deuterons, and thus there will be more helium production. But, as I have said, the more types of neutrinos you have, the faster the rate and the more neutrons there will be. One can see from the theory in detail how the ratio depends on this, and use its value to limit the number of types of neutrinos. The result is that the best fit is with three flavors, which is what our terrestrial experiments reveal—a remarkable result.

After they became too cold to interact rapidly, the neutrinos expanded freely with the universe. Today their temperature is slightly cooler that that of the photons, which is about 2.7 degrees above absolute zero. There are about 411 cosmic background microwave photons per cubic centimeter everywhere in the universe. These are what are detected. On the other hand, there are about 112 neutrinos of each flavor per cubic centimeter everywhere in the universe, and these cannot be detected. These neutrinos are what is called "hot dark matter" because of their energy. They cannot be captured by galaxies, which, as we shall see, is important in the search for the missing dark matter.

10

Squarks, Tachyons, and the Graviton

If we have learned anything from what has gone before, it is that theorists' speculations should be taken seriously—or at least considered.

In 1916, after he had created his theory of gravitation, Einstein made the following suggestion. On analogy with the fact that accelerated charges radiate electromagnetic waves, accelerated masses should radiate gravitational waves. In Einstein's theory these gravitational waves produce ripples in space-time. That such radiation exists has now received spectacular confirmation. In 1974 Joseph Hooton Taylor Jr. and Russell Hulse found a system of binary pulsars. These are massive stars that revolve around their common center of mass. Because of their mass, they should radiate gravitational waves with considerable intensity. This means that they should be continually losing energy and that therefore their orbits should be shrinking. Figure 20 shows how this progresses in time. With the change in orbit the period of revolution also changes, and

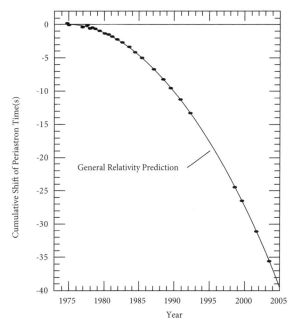

Figure 20. Orbital change due to gravitational radiation. Graph courtesy of Joel Weisberg, David Nice, and Joseph Taylor.

Einstein's theory tells us by how much. This prediction has been confirmed, which certainly means that gravitational waves exist. Nonetheless, it seems desirable to confirm this with terrestrial experiments. One of them is the Laser Interferometer Gravitational-Wave Observatory (LIGO). It consists of two long tubular arms in an L shape. A laser emits a beam that is split so that a portion runs down each arm and should arrive simultaneously at the ends. If a gravitational wave is incident on the device, it alters its length and changes the synchronism. It is this change that the experimenters have been looking for—so far without success.

With the creation of the quantum theory it was natural to try to fit gravitation into the mix. Hence theorists imagined a gravitational quantum, which acquired the name graviton. Properties of the graviton can be given sight unseen. It must be massless, because, like the electrostatic force, gravitation is long-range. Thus it must move with the speed of light. It must be electrically neutral, for charges are not exchanged in a gravitational interaction. Finally, it must have spin 2. This has to do with how gravity couples to matter in Einstein's theory. But does the graviton exist?

Freeman Dyson has given an argument that shows that single gravitons cannot be detected by the LIGO. This is because to overcome the quantum fluctuations that limit the accuracy of the measurement of the lengths, the device would have to be so massive that it would collapse into a black hole of its own making. Perhaps some other device might detect them, however, or perhaps, like the quark, they may be unobservable.

A number of years ago I attended a lecture that Feynman gave at Columbia University. His theme was to show, as he put it, how an ordinary guy such as himself—not an Einstein—might have discovered Einstein's theory of gravitation. He drew some diagrams showing graviton exchange and extracted Einstein's predictions without any mention of curved space and time. Quantum gravity was simply a theory like all the others. Then he drew diagrams with the quantum corrections. They were all infinite. Indeed, the number of infinities in the theory is infinite. The theory is not renormalizable—it makes no sense. This suggests that we take stock of where we are.

As we have seen, quantum electrodynamics allows us to make incredibly accurate calculations. One might compare the accuracy of calculating the electron magnetic moment to trying to measure the distance from Los Angeles to New York to within the width of a human hair. But while the infinities have been shoved under the rug, they are still there. The standard model, which includes the quarks and gluons, the leptons and the weak mesons W and Z, as well as quantum electrodynamics, is in the same boat. One can calculate anything one needs to, and one finds agreement with experiment. If the discovery of the Higgs boson is confirmed, then the theory is in that sense complete.

However, gravitation is in a different category. On one hand, there is Einstein's beautiful classical theory of general relativity and gravitation, which has passed every test put to it, from the anomalies in the orbit of the planet Mercury to gravitational radiation. On the other hand, attempts to quantize it using traditional methods have failed. There is no rug large enough under which to put the infinities. Something different is called for, and one popular candidate is what is known as string theory. Numerous books have been written at all levels on this, so here I am going to make only a few brief points.

The usual treatment of the quantum theory of fields—of which quantum electrodynamics is an example—supposes that the particles in the equations are points in space and time. A trajectory of such a particle is a one-dimensional line. This squeezing down of the extension of the particles has been suggested as the source of the infinities.

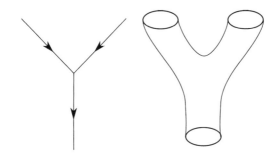

Figure 21. The "pants diagram."

Thus some physicists proposed to replace the point particles by tiny loops of string, making the trajectories into tubes. Figure 21 shows an example of what one of the founders of the theory, John Schwarz, calls the "pants diagram." On the left is the old theory and on the right the string version. On the right, the particles are represented by the closed loops at either end of the diagram.

These strings have tension, which would cause them to collapse except that Heisenberg's uncertainty principle won't allow it. But the strings can have excited modes that correspond to different particles. The first string theory applied only to bosons, which made it seem a bit academic. However, it was discovered that one of the bosons was a

massless, chargeless boson of spin 2 that was identified with the graviton. Hence there seemed to be a
road to a consistent theory of quantum gravity. But
clearly without fermions the theory had limited application. Then came the notion of supersymmetry.

The symmetries we are familiar with unite bosons with bosons and fermions with fermions. An
example is what is known as isotopic spin. Suppose
we lived in a world in which there were only the
strong interactions; all the other interactions were
turned off. In that world it would be plausible that
the neutron and proton were two states of the same
particle. For example, they would have the same
mass. The three pions would also collapse into a
degenerate triplet. This symmetry would manifest
itself in the strong interactions among these particles. Now imagine that we turned the other interactions back on. Manifestations of these relationships
would remain, for the symmetry breaking did not
destroy them completely. Prior to the work on
strings it was suggested that there might be super-
symmetries that related bosons and fermions,
though it was recognized that these would not be
discernible among the particles we know. There
would have to be a new array of superparticles that

would share the symmetry. These particles were given cute names such as squark and slepton. It was realized that if you added supersymmetry to string theory, you might come up with something that had a bearing on the real world.

There were two important consequences of adding supersymmetry to string theory, of which at least one was testable: that these new particles should show up in experiments. In many of the models they had masses that were attainable by existing accelerators. So far none has shown up, and some of the models have been ruled out. As far as I know, there is no real argument for a maximum mass of these objects, unlike in the Higgs case. One can always say that they are too massive to have been produced in the existing accelerators. I do not see any movement toward building still-larger accelerators, however; the Large Hadron Collider may be the end of the line. In that case the only hope of finding such particles is in outer space, using some satellite or other.

The second consequence, as far as I can tell, resists direct observation. All supersymmetric string theories that have any chance of accommodating what we know require extra dimensions. That is to say, they require more than three spatial dimensions, although as a rule there is still one time dimension. If we look at the history of trying to unify the forces, we can start in 1905 with Einstein's theory of special relativity. What Einstein recognized was that electricity and magnetism were two manifestations of the same force. Which manifestation we experience depends on our state of motion with respect to charged objects. At rest we may see a pure electric field, but once we are in motion a magnetic field appears. After Einstein created his general theory of relativity and gravitation it was natural to ask if these too could be unified. The first work on this subject was done by Theodor Kaluza in 1921. This was followed by the work of Oskar Klein in 1926. Klein said that to carry out his program he would need an extra spatial dimension, which we are not aware of because it is curled up in a tiny circle. This accomplishes the unification, albeit perhaps with some loss of plausibility. But now theorists are after much bigger game. They want a theory of everything—all the particles and all the interactions. The models that are now in favor have ten spatial dimensions and one time dimension. Seven of these spatial dimensions are not visible, at least not directly. I am reminded of what Robert Oppenheimer would say if someone came into his office in Princeton with a somewhat outré idea: "I'm glad that there are people thinking along those lines."

There are two mysteries that have in common the adjective *dark*—dark matter and dark energy. That may be all they have in common. Dark matter is matter that does not shine. It is probably composed of particles, which may even be the supersymmetric ones. Dark energy is distributed all over the cosmos uniformly and indeed is responsible for something like 72 percent of the mass-energy of the universe. I now want to discuss the evidence for these things and the most popular suggestions for the resolution of these mysteries.

The discovery of dark matter goes all the way back to the 1930s. Galaxies in clusters rotate. There is a connection between the kinetic energy of these rotations and the gravitational energy that keeps them from flying apart. This connection is empirically violated in some galaxies, however, in which the speed of rotation of the stars does not fall off as the distance from the center increases. It remains uniform out to large distances beyond the visible stars. This implies that there must be gravitating masses that are not visible. These invisible gravitating objects account for about 83 percent of the gravitating matter. There is other evidence, but this by itself is pretty convincing. Two things must be true of whatever this matter is: it must be quite massive, and its only significant interaction with anything must be gravitational. If there was an electromagnetic interaction, the stuff would shine. There is a suggestion that this matter might consist of weakly interacting massive particles with some kind of force that we have not seen before. I have noted earlier that they cannot be ordinary neutrinos, since these are too "hot" to be bound in galaxies. One suggestion is that they are "sterile" neutrinos, which are massive and do not have any other interactions that are not gravitational. Maybe the particles are, as I said, of the supersymmetric kind. We are, if I may say so, in the dark.

The history of dark energy goes back even further than that of dark matter. In 1917 Einstein published what was the first modern paper on cosmological theory, using his theory of general relativity and gravitation. He based it on two important assumptions that were consistent with what was known empirically at the time. The first was that the Milky Way galaxy represented the limits of the universe. There were no stars outside it. The second assumption was that the situation was static. There were "proper" motions among the stars, but on the average there was no expansion or contraction of the system as a whole. Einstein's general relativity theory has as its first approximation Newtonian gravity. But Newton encountered a paradox with a similar model, so it is not surprising that Einstein encountered the same paradox.

Suppose, Newton argued, that on the average the mass of the universe was spread over it with some density. Inevitably there would be fluctuations, if only because the matter was warm. Such a fluctuation would create a local increase in the mass somewhere. But this mass would attract more mass gravitationally, and on and on. The whole thing would collapse someplace. To avoid this, Newton assumed that the universe was infinite, so there was no place to single out.

Einstein had the same problem, but he resolved it by altering his theory. Einstein's theory is built on the idea that its equations should have the same form in all reference frames, including ones that accelerate. The special theory of relativity was special because it applied only to reference frames that were moving uniformly with respect to each other. But the equation that Einstein wrote down was not quite the most general one consistent with his assumption. He could add a constant term. He did not do this because there was no need, and if he had, the theory would not have morphed into Newton's. But now he needed to, to keep his universe static. This introduced a new constant, Λ—the cosmological constant. Einstein was free to choose the sign and magnitude, and he chose them so as to cancel out the gravitational force in his cosmology. His universe was static.

In the meantime, a Russian polymath named Aleksander Friedmann had been considering in the context of Einstein's original theory how the scale factor I discussed before might evolve in time. He found that this depended on the density of matter. You could have continuous expansion, a static universe, or one that contracts. You might even imagine one that is cyclic, with a new Big Bang recurring periodically. Einstein's first reaction to this was to say that it was wrong. Subsequently he said that it was right but irrelevant since the universe was static. Then came Edwin Hubble.

Hubble was an American astronomer who spent most of his professional life at the Mount Wilson Observatory near Pasadena, California. He was able to take advantage of what was then the largest telescope in the world. First he confirmed that there were galaxies outside the Milky Way. Then in the late 1920s he discovered that the universe was expanding. He was able to determine the distance to some of the galaxies and observed that the farther away they were, the more their light was shifted to the red end of the spectrum—a Doppler shift. This meant that they were receding with a speed that increased with the distance from us. It took some time to associate this discovery with Friedmann's universe. Hubble's discovery has been confirmed over and over again. Until 1998 all the workers in this field would have told you that the universe was decelerating because the expansion was being pulled back by gravity. But in 1998 it was announced that observations on some supernovae showed that in fact the universe is now accelerating and has been doing so for the last 5 billion years. With

Hubble's discovery Einstein had stated that his introduction of the cosmological constant was a blunder. But now it seems as if that constant might be back.

Indeed, one of the first explanations of these new data was the reintroduction of the cosmological constant. Things had changed since Einstein's time, however. He had the luxury of choosing the constant to suit his needs. But now it had to have an explanation. An explanation can be found in the quantum theory of fields. It bears a family resemblance to how the Higgs boson generates mass, but in this case there is a terrible problem: the field theories predict a constant that is about 120 orders of magnitude too large. One of the lessons of this book is that if you find a numerical discrepancy of this size, there must be an explanation, often in terms of a new symmetry. Theorists have been looking in vain for such an explanation, and once again we are in the dark.

*

The last particle I want to discuss is the tachyon. It was so named by the late Gerald Feinberg, who was a professor at Columbia. *Tachus* is Greek for "speedy," and if these particles are nothing else, they are certainly that. They move faster than the speed of light. Indeed, they can never move slower than the speed of light. The speed of light in a vacuum has a special role in relativity, in that every observer, no matter in what state of relative motion, measures the same speed. To put it another way, you can't catch up with a photon. Einstein referred to this as the principle of constancy. When I first learned about it from Philipp Frank I was both amazed and puzzled. Part of my puzzlement involved how Einstein had ever thought of this. It could not have been from experiments on moving bodies and light, since any moving body that he had available moved at a crawl. For some time I did not understand Professor Frank's answer, which was that the ratio of the physical dimensions of an electric field to a magnetic field is a constant, with the dimensions of a velocity that turns out to be the speed of light.

Terrible things happen to the equations of relativity when you try to make a massive particle move with the speed of light. On the other hand, if the particle always moves faster than light, then you can get away with it, but at a price. If you can use these tachyons to carry information, then you can reverse cause and effect if you are not careful. If in one system event A follows event B, then you can find another in which the order is reversed. It does not take much imagination to invent malign scenarios of all kinds. Feinberg got around this by noting that you can arrange tachyons so that the emission of a tachyon backward in time can be replaced by the absorption of an antitachyon going forward in time. He was so taken by all of this that he even participated in an experiment to try to find tachyons and antitachyons, though without success. On the other hand, in late 2011 a group of Italian experimenters claimed that neutrinos produced at CERN were moving faster than light. When I heard about this I remembered Einstein's reaction in the 1930s when he was told that an experiment had disproved the principle of relativity. He remarked, "Raffiniert ist der Herrgott, aber boshaft ist Er nicht"—"God is sophisticated but not malicious." That experiment was proven wrong; so was this one.

L'Envoi

Good mystery stories have a neat plot with a tidy
ending. This story is more like a series of nested
Russian dolls: inside each one there is another. I do
not believe we are at the end.

Appendix I
Accelerators and Detectors

The purpose of this appendix is to bring together and amplify a theme that has recurred throughout this book. To probe deeper and deeper into the interior of matter, the tools used have become ever more sophisticated. I began working in this field a half century ago, and when I began this sort of evolution would have been disregarded as wildly speculative science fiction. I wrote my Ph.D. thesis in the mid-1950s. The largest part of the work that I had to do was numerical. This was done with a Marchant calculator, which was electromechanical and performed its arithmetic operations by moving gears. If you divided a number by zero, the machine kept grinding until the gears burned up. It took me literally months to carry out this work. When I had finished it I discovered that a fellow at MIT had done a rather similar thesis. He too had numerical work to do. But he had had the good sense to cultivate a group at MIT that was building the latest version of the Whirlwind computer, which was the state of the art. It used vacuum tubes and could process about 40,000 instructions a second. As I recall, it took about an afternoon to do his work. Of course, by present-day standards this machine was a dinosaur. Your laptop can probably process a hundred million instructions a second.

In experimental elementary particle physics, computers are used for much more than numerics. Perhaps the most important use is in pattern recognition. In my day at the Harvard Cyclotron there were a few women employed to scan tracks made on photos and sort out the interesting events. I suppose that there might have been substantially less than a hundred photos involved in any experiment. Later, as I will discuss, the Gargamelle bubble chamber at CERN, where an important discovery was made, produced more than 100,000 tracks that had to be scanned. This was done by computer.

In the body of the text I have avoided introducing the energy and mass units that are standard for this subject. I did this because these units are so far outside our daily experience. Instead I compared masses of particles without ever telling you what these masses are. But in this appendix I want to bite the bullet and introduce these units. I will begin with a unit we all know—the watt. The watt is a unit of power, energy per unit of time—per second, for example. When your meter is read it gives numbers in kilowatt-hours. This is the electrical energy you have purchased during the period between meter readings. The energy unit that is used in defining the watt is the joule, named after the nineteenth-century British physicist James Prescott Joule. The watt is named after James Watt, who lived a bit before Joule. One watt is one joule per second. A kilowatt-hour is 3.6×10^6 joules.

Voltage is the measure of the work you have to do to move a unit of electric charge between two points in an electric field. Equivalently, it is the energy gained if you reverse this process. For our subject the important unit is the electron volt (eV). This is the energy gained if an electron moves across a voltage drop of one volt. It has a unit of energy—for example, the joule. But in terms of joules it is an absurdly small number. One electron volt equals 1.6/10,000,000,000,000,000,000 joules (1.6×10^{-19} joules). That is why I did not introduce it before. It makes sense only in the context of elementary particle or atomic physics, where things are naturally measured in these units. We will run into meV (which is equivalent to 10^{-3} eV), keV (10^3 eV), MeV (10^6 eV), GeV (10^9 eV), and TeV (10^{12} eV). (In my day a GeV was called a BeV, so the Bevatron was designed to accelerate protons to this kind of energy.)

It is very convenient to use Einstein's $E=mc^2$ to convert masses into energies. Let me give an example so you will see why. In grams the mass of the electron is 9.109383×10^{-31} kilograms. The speed of light in meters per second is 299,792,458 m/s. You must square this and multiply by the mass in kilograms to get joules. Then you use the conversion of joules to electron volts to get the mass-energy in electron volts. If you do this, you will find that the mass-energy of the electron is 0.510998928 MeV. It is much easier to keep track of a half an MeV than the absurd number in grams. In the future I will simply, as is customary in our business, refer to mass-energy as the "mass."

In this spirit the mass of the proton is 938.272046 MeV, while the mass of the neutron is 939.565379 MeV. You will notice that the neutron mass exceeds that of the proton by enough that the decay into an electron is allowed by the conservation of energy. Since the positive and negative muons are antiparticles of each other, they have the same mass, which is 105.6583715 MeV. The positive and negative pions are also antiparticles of each other, so they have the same mass, which is 139.57018 MeV. The neutral pion is its own antiparticle and has a mass of 134.9766 MeV. It is slightly less massive. The positive and negative K-mesons are also antiparticles of each other and have a mass of 493.677 MeV. The masses of the neutral kaons are tricky. As we have seen, the neutral kaons are not antiparticles of each other. The states that have definite mass are not the states that are created but the states that have evolved in time. There is a small mass difference between these states, and if we ignore it, then the mass of the neutral kaon is 497.614 MeV.

I do not intend to run through the whole list; the reader can refer to the tables in the text. But I want to give the mass of at least one nonmesonic strange particle, the Λ^0. Its mass is 1115.68 MeV. Finally I will give the masses of the W and Z mesons. The charged W particles are antiparticles of each other and have a mass of 80.399 GeV. (Note that we are now in the GeV range.) The mass of the Z is 91.1876 GeV. All the masses that I have given have an uncertainty attached to them, which I have not included.

✳

It is not my intention to write anything like a complete history of particle accelerators. That would require another book. Instead I will make a few points that will give the general flavor. I would argue that the first particle accelerators were the cathode ray tubes used by J. J. Thomson to discover the electron. Remember that these were evacuated glass tubes with the negatively charged cathode at one end and the positively charged anode at the other. When the cathode is heated electrons are emitted, and these are accelerated by electric fields, so they strike the anode with an enhanced velocity. Recall that Thomson measured the change in temperature of the anode due to the absorption of this kinetic energy.

In Thomson's experiments a potential difference of a few hundred volts was established between the cathode and the anode. By the very early 1930s it was possible to produce a few hundred kilovolts to use in the acceleration process. One of the devices was invented by an American engineer named Robert Van de Graff. This consisted of a metal sphere that could be charged up to millions of volts. It is not clear what physics Van de Graff had in mind. But it is very clear what physics the two British physicists John Cockcroft and Ernest Walton had in mind. They wanted to accelerate protons to an energy at which quantum mechanics predicted they would penetrate nuclei. There is an electrical energy barrier that repels the protons, but quantum mechanics tells us that it can be penetrated by tunneling through it. Cockcroft and Walton built up a potential difference of some 800 kilovolts, and in 1932 they produced protons that were energetic enough to penetrate the lithium nucleus, producing two alpha particles. All of these devices were linear accelerators. This was about to change.

In 1929 the American physicist Ernest Lawrence read a paper by the Norwegian engineer Rolf Wideröe. To say he "read" the paper is a bit of a euphemism, as Lawrence could not read German, but he got the gist by looking at a diagram. Wideröe was suggesting that accelerating could be achieved in stages by alternating the polarity of the electric fields. Lawrence recruited a young electronics expert who built a working model that in 1931 managed to accelerate mercury ions to an energy of 1 million electron volts. This was still a linear accelerator, but about this time Lawrence had an idea of genius that changed everything. Why not make the charged particles move in circles, which could be done by applying a magnetic field at right angles to the orbital plane? After a semicircle one can apply a suitable electric field to give the particles a little boost of speed. After the next semicircle the field has to be reversed to give it a new boost. Lawrence's observation of genius was that the time it took for any circular orbit under these conditions was the same for all orbits. As the orbits got larger the particle speeded up in just such a way as to make the time the same. This vastly simplified the electronics.

The first successful model was built by one of Lawrence's postdoctoral students, M. Stanley Livingston. The first one that Livingston built was about four and a half inches in diameter. He was able to use vacuum tube technology to supply the periodic changes in the electric field. It cost about $25 to build. It may have looked like a toy, but it produced protons of 80 keV. Lawrence kept improving and expanding his model, but the principles remained the same. This included the Harvard cyclotron. By the time I left in 1957 the energy had been boosted to about 165 MeV, too low to do any real pion physics, but it facilitated some precision nuclear physics experiments, and several generations of physicists got their degrees using it.

A 100 MeV proton moves with a speed that is somewhat less than half the speed of light. This means that the effects of the theory of relativity, which depend on the square of this number, can largely be ignored. However, once you are interested in protons that have 1 billion electron volts of kinetic energy, relativity can no longer be ignored. A particle that moves past you with a speed approaching the speed of light has an effective mass that is larger than the mass of the same particle when brought to rest. This changes everything when it comes to accelerator design. You can no longer use a magnetic field of a fixed strength to guide the particles nor a fixed-frequency electric field. These quantities must be synchronized with the increasing speeds to take into account the changes in mass. The new generation of accelerators that were designed with these effects in mind were appropriately called synchrotrons.

The first of these machines to come on line was the Cosmotron at Brookhaven, in the building where I used to play the trumpet when the machine was not running. It had a seventy-five-foot-diameter ring and reached its full proton beam energy of 3.3 GeV in January 1953. It ran until 1968. The next was the Bevatron in Berkeley, which went online in 1954 producing protons of 6.2 GeV—just enough energy to make antiprotons. It was decommissioned in 1970.

I do not intend to run through the whole list, although it is not that long. Each machine was very expensive—the Cosmotron cost $8 million in 1950s dollars—and they took years to build. The step to the next-level TeV accelerators took new technology. For example, the Tevatron at Fermilab, which went online in 1983 and was finally shut down for lack of funds in 2011, took advantage of superconducting magnets, which replaced the earlier iron magnets. It produced protons of 980 GeV.

An entirely new kind of accelerator was introduced in the 1980s. The prime example was the Large Electron Positron (LEP) Collider, which went online at CERN in 1989. Electrons and positrons were injected into a ring that was 27 kilometers in circumference. These had been preaccelerated in a linear accelerator. The two beams went in opposite directions around the ring. The beams were focused so that they collided in four places around the ring where detectors were in place. By the time the machine was shut down in 2000 it had produced colliding beams of some 209 GeV. Its successor, the Large Hadron Collider, collides protons and is expected to reach an individual beam energy of 7 TeV when it is at full strength. These machines are all marvels of technology and bear about as much resemblance to Lawrence's first cyclotron as the drawings of Leonardo's flying machines have to a jet airliner.

Along with the advance in accelerator technology there was a parallel advance in detector technology. It is interesting to recall Rutherford's experiment in which the atomic nucleus was discovered. Alpha particles—helium nuclei—produced in the decay of radon were made to collide with thin metal foils. One usually reads that these were gold foils, but several other metals were used as well. After the alpha particle penetrated the foil it struck a zinc sulfide screen, causing a flash of light. Rutherford's assistants, Geiger and Marsden, sat in a darkened room an hour or so before the experiment started, to accustom their eyes. They then looked at the flashes through a microscope. That is collecting data retail. In the 1930s, as I have mentioned in the text, Wilson cloud chambers were used to detect the tracks of charged particles produced in cosmic rays. This way both the positron and the muon were discovered. After the war photographic emulsions were used to observe the strange particles in cosmic rays. This was also collecting data retail. Once accelerators such as the Cosmotron began producing beams of these particles, a new method of detection was called for, and indeed, just at this time one was found. This was the invention of the bubble chamber by the physicist Donald Glaser.

To understand the workings of the bubble chamber, think of the physics of boiling. Say you put water in a kettle and turn on the stove. As you increase the heat, the bonds that hold the liquid together begin to give way and the water begins to vaporize. Small water vapor bubbles are formed. If you look at the forces that act on the bubbles, you can identify three. Inside the bubble there is a vapor pressure that acts to try to expand the bubble. Opposing this is the surface tension of the bubble and the pressure from the ambient environment. The temperature at which these forces equalize is called the boiling point. The bubbles grow in size and are visible as they rise to the surface. The boiling point depends on the environment in which the heating is taking place. If you have ever camped at high altitude, you will have observed how hard it is to boil an egg. There is less atmosphere pressing down on the water and the temperature of the boiling point is lowered. What Glaser made use of is the fact that some liquids can be superheated—they can be maintained at least temporarily at temperatures above their nominal boiling points. In the bubble chamber the pressure on the liquid is reduced just as the particles to be detected enter the chamber. Such a particle deposits its energy on a bubble, causing it to expand and become visible. What one sees is a track made by these particles. An early application was a liquid hydrogen bubble chamber at the Bevatron that was used to detect the antiproton.

The mother of all bubble chambers was the Gargamelle. (Gargamelle, invented by Rabelais, was the giantess who was the mother of Gargantua.) The Gargamelle, which went into operation at CERN in 1970, was filled with Freon, a relatively dense liquid. It was designed to detect neutrinos that were produced by the Proton Synchrotron (PS), which was the predecessor to the LEP. The PS accelerated protons to 25 GeV, producing massive numbers of pions, which produced neutrinos when they decayed. The great discovery made at the Gargamelle took place in 1973, when researchers produced evidence for what became known as the weak neutral current. This was prior to the actual discovery of the W and Z mesons, but it was essential for the standard model that these exist. The W mesons carry electric charge and mediate processes such as the decay of the neutron, where it decays into a positively charged proton. The Z meson mediates weak processes in which the charge is not changed. For example, a neutrino can collide with an electron, transferring its momentum to it. This type of event— the appearance of energetic electrons—is what the Gargamelle was looking for. After 83,000 events were analyzed there were 102 neutral current events discovered.

The Gargamelle represented the high-water mark of the bubble chambers. This technology had some disadvantages, including the fact that the superheated phase had to be ready at precisely the time of the particle collisions, which made it impossible to detect the reactions of very short-lived particles. Bubble chambers have been replaced by detectors such as the wire chamber, in which one or more electric wires are suspended in a gas. When an ionizing particle enters the chamber it produces a cascade of charged particles, which gather on the wire, making an electric current. The magnitude of this current is a measure of the energy of the particle being detected.

The LHC illustrates every modern detector type. The circular tunnel in which they are contained has a circumference of 17 miles. The tunnel at some points is more than 500 feet belowground. There are six detectors sited around the tunnel. The machine, the tunnels, and the detectors cost about $9 billion to build. The ATLAS detector, for example, is about half as big as the Notre Dame Cathedral in Paris and weighs more than the Eiffel Tower. There are 3,000 physicists associated with it. We are in a realm of experimental physics different from anything I could ever have imagined.

Appendix 2
Grand Unification

In the fall of 1957 I arrived at the Institute for Advanced Study in Princeton, New Jersey. I was told that the director, Robert Oppenheimer, wanted to see me. I could not imagine why, but the first thing he asked was, "What is new and firm in physics?" I was rendered mute. Whatever I knew that was new was not firm, and vice versa. Fortunately Oppenheimer's phone rang, so I could escape without answering the question. In most of this book I have emphasized what is firm. A reader who is encountering the material for the first time is, I think, better off staying away from speculations that have no real experimental base and which may turn out to be wrong. But in this appendix I want to discuss some speculations. I will try to keep the discussion at the same level as the body of the text, but it may become slightly more technical.

First I want to deal with the notion of coupling constant. We have seen the simplest Feynman diagram that represents the collision of two electrons

(Figure 8). But what I did not indicate in that diagram was that at each place in the diagram where a virtual photon is emitted or absorbed there should be a factor of e—the electric charge that measures the strength of the interaction. This diagram involves two such places, so we would say that it is of order e^2. But the e that enters here has the curious dimensions of energy times distance. Thus one cannot say if it is large or small. If one looks more closely, one finds that what actually enters the diagram is $e^2/\hbar c$, where $\hbar$ is Planck's constant divided by 2π. This number is dimensionless and is known as α. It is approximately 1/137. Corrections to this diagram involve higher powers of α and hence should be small. As I discussed in the text, they seemed to be infinite until the whole post–World War II renormalization program showed how to deal with them.

We have seen that there are four forces we have to deal with. The other three are also characterized by coupling constants, which are given approximately by:

α_{strong}	1
$\alpha_{electromagnetic}$	1/137

$$\alpha_{weak} \qquad 10^{-6}$$
$$\alpha_{gravity} \qquad 10^{-39}$$

The fact that these constants have such diverse values is one reason many theorists are discontented with the standard model, which is what I have been explaining in the body of the book. The standard model with its quark color dynamics and its unification of the electromagnetic and weak interactions is certainly one of the most successful scientific theories ever invented. It answers every question put to it. In a way this is a pity, for if it broke down somewhere, we might learn something. What bothers these theorists is its semiempirical basis and its lack of unification. To make the comparison with experiment, one must insert values of things such as the coupling constants, which in turn one takes from experiment. Moreover, there is no explanation for these constants. For example, why is the magnitude of the proton charge the same as that of the electron even though the electron was probably produced by the Big Bang while the proton was created from bound quarks later? We could pose the charge question in terms of quarks by asking why these have charges that are simple fractions of the proton charge.

The unification that has been tried can be broken into two parts—kinematic and dynamic. In the kinematic part one enlarges the symmetry structure of the theory so that particles such as quarks and electrons fall under the same rubric. This helps to explain why these particles have related charges, although it does not explain the value of, say, $\alpha_{electromagnetic}$. It also does not explain why the coupling constants for the different forces are so very different. This, as I will now attempt to argue, is a dynamical problem. Let us begin with the electromagnetic case.

In the case of the electron-electron scattering depicted in the Feynman diagram, we simply put in the value of the electron charge that we observe in experiments such as this. We can call this α_{obs}. We might think that we can use this value for any collisions no matter what the energy of the electrons. But quantum mechanics tells us that this is not the case. Let us consider one of the electrons. These considerations apply mutatis mutandis to the other. The space around this electron is a roiling sea of virtual electron-positron pairs. These arise out of the vacuum violating the conservation of energy but for a time short enough to conform with the

Heisenberg uncertainty principle relating energy and time. The violation of energy conservation is roughly of the order of the electron rest mass energy mc^2, so the allowed time is of the order of $\hbar/mc^2$. If we assume that these virtual particles move with the speed of light, then during this time they will move a distance of the order of $\hbar/mc$, the Compton wavelength of the electron, which is about 4×10^{-13} meters. If the incident electron is not energetic enough to penetrate this distance, the charge it sees is affected by the presence of these virtual pairs. The positrons shield the bare charge, so the observed charge is lower by an amount that depends on the energy. It is only when the energy is high enough that the bare charge is relevant. Similar considerations apply to the coupling constants for the other forces. The weak force produces fluctuations involving the weak mesons. The quark force is somewhat different because the binding increases the farther away the quarks are from each other. Figure 22 is a graph that shows in a model calculation how these charges vary with energy.

Note that the coupling constants merge at about 10^{16} GeV. Gravitation does not quite work, so I have left it out. To make it work at all, one must go into

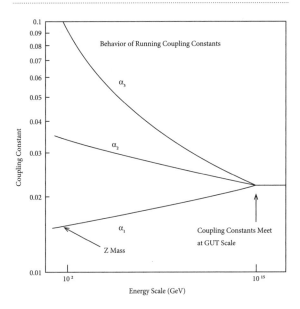

Figure 22. The running coupling constants.

higher dimensions and introduce a realm of new particles, at least some of which should be accessible at the Large Hadron Collider. If none is observed, we may have to go back to the drawing board.

Neutrino Oscillations

Bruno Pontecorvo was born in Pisa in 1913. He was something of a prodigy, and at age eighteen he was admitted to Fermi's course in Rome. He then became a member of Fermi's experimental team, where he remained until 1936, when he moved to Paris to join the group of Irène and Frédéric Joliot-Curie. Joliot-Curie had very leftish political ideas. In fact, during the war, when he took part in the Resistance, he joined the Communist Party. Pontecorvo had had no political interests, but, influenced by Joliot, he adopted some of the latter's ideas. In 1938 the racial laws were introduced in Italy, and Pontecorvo, who was Jewish, could not return. He just escaped the German occupation of Paris and emigrated to the United States, where he found employment with an oil company in Oklahoma. He introduced neutron bombardment as a method of measuring the geology of the boreholes as they were being drilled. In 1943 he was asked to join the reactor project in Canada, and in 1948 he became a British citizen and worked on the British atomic weapons program. On August 30, 1950, on a visit to Stockholm, he was helped by Soviet agents to defect to Russia. There was great concern that he might transmit nuclear secrets, but apparently he did not have any essential information. There is no evidence that he worked on the Russian nuclear weapons program, and indeed, he wrote articles urging the end of nuclear weapons. He died in 1993 in Dubna.

One of his interests was the neutrino. In fact, he was the person who suggested the method of detecting antineutrinos that was used by Cowan and Reines in their discovery. To remind you, the antineutrino—if it is of the electron type—will produce a positron in the reaction $\nu^c + p \rightarrow e^+ + n$. The positron rapidly annihilates with an atomic electron, producing two gamma rays, and a short time later the neutron has a detectable nuclear reaction. This sequence is what is observed in reactors. By 1957 the idea that there might be two types of neutrinos— muon and electron—had a certain amount of traction, although the experiment that confirmed this was done five years later. As I recall, almost everyone assumed that these two neutrinos were massless.

One attractive feature of this was that it seemed to give an explanation for the nonconservation of parity. Recall that a massless neutrino has a handedness or chirality. This can be explained if the interactions that produce these neutrinos have a handedness symmetry, which is possible only if the neutrinos are massless. But one can readily persuade oneself that these interactions violate parity conservation. Hence there seemed to be an elegant connection among these ideas. But sometimes in physics elegance is a form of fool's gold.

In 1955 Gell-Mann and Pais published their paper on neutral kaons. It will be recalled that in the strangeness-conserving strong interactions, all the production is of a neutral K-meson, which has a definite strangeness, or its antiparticle, which has the opposite strangeness. But these two objects can convert into each other by weak strangeness-violating interactions. Hence if, say, a neutral kaon is produced at some time, then at a later time it will transform itself into a mixture of itself and its antiparticle. This mixture evolves in time—that is, it oscillates—and this is something that can be studied experimentally. In 1957 Pontecorvo asked whether the same sort of thing might happen with the neutrino. He was thinking at first of only one type of neutrino, but after the 1962 discovery of the two neutrino types he used a generalization of this idea. This was a very bold speculation, as in order for this oscillation effect to take place, at least one of the neutrinos had to have a mass. The parameter that determines the oscillation is proportional to $\Delta m^2_{e\mu} = m^2_{\nu e} - m^2_{\nu\mu}$, the difference of the squares of the masses. If both masses are the same, the parameter vanishes and so does the effect. At the time the prejudice was for zero mass, so I do not think this idea had a wide following. In 1969 Pontecorvo followed this up in a paper he wrote with the Russian physicist Vladimir Gribov. By this time it was known that something was wrong with the theory about neutrinos produced in the Sun: too few of them arrived on the Earth. Gribov and Pontecorvo proposed an explanation in terms of neutrino oscillations. This was certainly a possibility, but I believe most physicists thought there was probably something wrong with the solar model.

The first real experimental evidence for neutrino oscillations was announced in 1998. This was an experiment done in the so-called Super-Kamiokande observatory, which is located some 3,000 feet belowground in a mine under Mount Kamioka in Japan. Situated there is a very large tank of very pure water. The idea is that if an energetic neutrino ejects a charged particle, that particle can move faster than the speed of light in water. This produces a special kind of radiation known as Cherenkov radiation, which is what is observed. The detector is sensitive to both muons and electrons and hence should be able to detect both electron and muon neutrinos. The kind of neutrino it was looking for is known as "atmospheric." These are the secondary results of cosmic radiation. The primary protons produce pions, which decay into a muon neutrino and a muon. The muon decays into an electron, an electron neutrino, and a muon neutrino. Hence the Super-Kamiokande detector should see neutrinos of both kinds. It did, but the ratio was wrong: there were too few muon neutrinos.

By this time it was known that there were three kinds of neutrinos, the third being the tau neutrino. Hence there were three kinds of possible oscillations. The muon-electron neutrino oscillation was too small to account for the effect. Hence it had to be the muon-tau neutrino oscillation. The tau is too heavy to be produced by this neutrino; hence if the idea is right, a certain number of muon neutrinos are put out of action. A confirmation is obtained by observing how the effect depends on the direction from which the neutrino is coming. Some neutrinos pass through thousands of miles of the Earth before they get to the detector, and others go only a short distance. The effect of the oscillation increases with the distance traveled by the neutrinos, and this directional effect was observed. There is no doubt that this was an observation of neutrino oscillations.

Since 1998 a variety of experiments have been carried out. There was seen solid evidence for oscillations between the electron neutrino and the muon neutrino as well as between the muon and tau neutrinos. The mass square differences for these neutrinos are comparable. What was missing was evidence for oscillations between the electron and tau neutrinos. This was produced in the spring of 2012 by reactor experiments done separately by two groups in Asia. Both of these experiments involve several reactors. There is a Chinese-American collaboration and a South Korean experiment. These experiments agree with each other and show that there is an oscillation effect that is smaller than that of the other two. These experiments only measure neutrino mass square differences, so the actual neutrino masses are not known. Also what is not known is the origin of these masses, although some theorists think that it may be a Higgs mechanism. It is possible that there is another family of extremely massive neutrinos that are conjugate to the ones we know but which have no interactions except gravitational. Possibly these objects are the dark matter.

Acknowledgments

I have profited from discussions, some electronic, with many of my physics colleagues. They include Elihu Abrahams, Steve Adler, Luis Alvarez-Gaume, Tom Appelquist, Frank Close, Kyle Cranmer, Freeman Dyson, John Ellis, Ken Ford, Howard Haber, Peter Kaus, Pierre Ramond, Jon Rosner, and Lincoln Wolfenstein. I am grateful to the *American Scientist* and Harvard University Press for their interest.

Index

accelerators. *See* particle accelerators

ace particles, 99

alpha decay, 30

alpha particles: discovery, 11, 12; plutonium decay, 31; Rutherford's experiment, 103, 181–182

Anderson, Carl, 71, 81–82

angular momentum: atom, 21; silver atom, 20; spin, 19

anode, 176

"anomalous" Zeeman effect, 21, 25

antideuterons, 77

antielectrons, 42

antikaon, 90t–91t, 175

antimatter, 71, 74

antineutrinos, 39, 71, 77, 193

antineutrons, 75–76

antiparticles: about, 70–77; of baryons, 91; charm-anticharm pair, 118; cosmology, 147; of deuterons, 77; of electrons, 42; Fermi on, 90–92; of K-mesons, 76, 175; of kaons, 175; of mesons, 86, 92; of neutrinos, 39, 71, 77, 193; of neutrons, 75–76; of photons, 71, 75; of pi-mesons, 71, 75; properties, 175; of protons, 71, 72, 74–77, 183; of quarks, 147

antiprotons: creation, 76; decay, 76; detection, 71, 72, 74, 75, 77, 183

antiquarks, 147

associated production, 89, 92

ATLAS detector, 141, 184

"atmospheric" neutrinos, 196

atomic nucleus, 62, 133, 181

atomic spectra, 21, 124

atomic structure: nuclear atomic model, 16; "plum pudding model," 11; Rutherford model, 16

atoms: angular momentum, 21; nature of, 10–11

Bahcall, John, 43

Balmer, Johann Jakob, 17

Balmer series, 17

baryon decuplet, 111–112

baryon number, 76, 88, 91–92, 114

baryons: antiparticle, 91; 125; quark content, 104, 107, and quarks, 104, 107, 120, 148